Nonparametric Statistical Methods Using R

Chapman & Hall/CRC
The R Series

Series Editors

John M. Chambers
Department of Statistics
Stanford University
Stanford, California, USA

Torsten Hothorn
Division of Biostatistics
University of Zurich
Switzerland

Duncan Temple Lang
Department of Statistics
University of California, Davis
Davis, California, USA

Hadley Wickham
RStudio
Boston, Massachusetts, USA

Aims and Scope

This book series reflects the recent rapid growth in the development and application of R, the programming language and software environment for statistical computing and graphics. R is now widely used in academic research, education, and industry. It is constantly growing, with new versions of the core software released regularly and more than 5,000 packages available. It is difficult for the documentation to keep pace with the expansion of the software, and this vital book series provides a forum for the publication of books covering many aspects of the development and application of R.

The scope of the series is wide, covering three main threads:

- Applications of R to specific disciplines such as biology, epidemiology, genetics, engineering, finance, and the social sciences.
- Using R for the study of topics of statistical methodology, such as linear and mixed modeling, time series, Bayesian methods, and missing data.
- The development of R, including programming, building packages, and graphics.

The books will appeal to programmers and developers of R software, as well as applied statisticians and data analysts in many fields. The books will feature detailed worked examples and R code fully integrated into the text, ensuring their usefulness to researchers, practitioners and students.

Published Titles

Stated Preference Methods Using R, *Hideo Aizaki, Tomoaki Nakatani, and Kazuo Sato*

Using R for Numerical Analysis in Science and Engineering, *Victor A. Bloomfield*

Event History Analysis with R, *Göran Broström*

Computational Actuarial Science with R, *Arthur Charpentier*

Statistical Computing in C++ and R, *Randall L. Eubank and Ana Kupresanin*

Reproducible Research with R and RStudio, *Christopher Gandrud*

Introduction to Scientific Programming and Simulation Using R, Second Edition, *Owen Jones, Robert Maillardet, and Andrew Robinson*

Nonparametric Statistical Methods Using R, *John Kloke and Joseph W. McKean*

Displaying Time Series, Spatial, and Space-Time Data with R, *Oscar Perpiñán Lamigueiro*

Programming Graphical User Interfaces with R, *Michael F. Lawrence and John Verzani*

Analyzing Sensory Data with R, *Sébastien Lê and Theirry Worch*

Analyzing Baseball Data with R, *Max Marchi and Jim Albert*

Growth Curve Analysis and Visualization Using R, *Daniel Mirman*

R Graphics, Second Edition, *Paul Murrell*

Multiple Factor Analysis by Example Using R, *Jérôme Pagès*

Customer and Business Analytics: Applied Data Mining for Business Decision Making Using R, *Daniel S. Putler and Robert E. Krider*

Implementing Reproducible Research, *Victoria Stodden, Friedrich Leisch, and Roger D. Peng*

Using R for Introductory Statistics, Second Edition, *John Verzani*

Advanced R, *Hadley Wickham*

Dynamic Documents with R and knitr, *Yihui Xie*

Nonparametric Statistical Methods Using R

John Kloke

University of Wisconsin
Madison, WI, USA

Joseph W. McKean

Western Michigan University
Kalamazoo, MI, USA

CRC Press
Taylor & Francis Group
Boca Raton London New York

CRC Press is an imprint of the
Taylor & Francis Group, an **informa** business

A CHAPMAN & HALL BOOK

To Erica and Marge

Contents

Preface

Nonparametric statistical methods for simple one- and two-sample problems have been used for many years; see, for instance, Wilcoxon (1945). In addition to being robust, when first developed, these methods were quick to compute by hand compared to traditional procedures. It came as a pleasant surprise in the early 1960s, that these methods were also highly efficient relative to the traditional t-tests; see Hodges and Lehmann (1963).

Beginning in the 1970s, a complete inference for general linear models developed, which generalizes these simple nonparametric methods. Hence, this linear model inference is referred to collectively as rank-based methods. This inference includes the fitting of general linear models, diagnostics to check the quality of the fits, estimation of regression parameters and standard errors, and tests of general linear hypotheses. Details of this robust inference can be found in Chapters 3–5 of Hettmansperger and McKean (2011) and Chapter 9 of Hollander and Wolfe (1999). Traditional methods for linear models are based on least squares fits; that is, the fit which minimizes the Euclidean distance between the vector of responses and the full model space as set forth by the design. To obtain the robust rank-based inference another norm is substituted for the Euclidean norm. Hence, the geometry and interpretation remain essentially the same as in the least squares case. Further, these robust procedures inherit the high efficiency of simple Wilcoxon tests. These procedures are robust to outliers in the response space and a simple weighting scheme yields robust inference to outliers in design space. Based on the knowledge of the underlying distribution of the random errors, the robust analysis can be optimized. It attains full efficiency if the form of the error distribution is known.

This book can be used as a primary text or a supplement for several levels of statistics courses. The topics discussed in Chapters 1 through 5 or 6 can serve as a textbook for an applied course in nonparametrics at the undergraduate or graduate level. Chapters 7 and 8 contain more advanced material and may supplement a course based on interests of the class. For continuity, we have included some advanced material in Chapters 1-6 and these sections are flagged with a star (*). The entire book could serve as a supplemental book for a course in robust nonparametric procedures. One of the authors has used parts of this book in an applied nonparametrics course as well as a graduate course in robust statistics for the last several years. This book also serves as a handbook for the researcher wishing to implement nonparametric and rank-based methods in practice.

This book covers rank-based estimation and inference for models ranging from simple location models to general linear and nonlinear models for uncorrelated and correlated responses. Computation using the statistical software system R (R Development Core Team 2010) is covered. Our discussion of methods is amply illustrated with real and simulated data using R. To compute the rank-based inference for general linear models, we use the R package Rfit of Kloke and McKean (2012). For technical details of rank-based methods we refer the reader to Hettmansperger and McKean (2011); our book emphasizes applications and statistical computation of rank-based methods.

A brief outline of the book follows. The initial chapter is a brief overview of the R language. In Chapter 2, we present some basic statistical nonparametric methods, such as the one-sample sign and signed-rank Wilcoxon procedures, a brief discussion of the bootstrap, and χ^2 contingency table methods. In Chapter 3, we discuss nonparametric methods for the two-sample problem. This is a simple statistical setting in which we briefly present the topics of robustness, efficiency, and optimization. Most of our discussion involves Wilcoxon procedures but procedures based on general scores (including normal and Winsorized Wilcoxon scores) are introduced. Hogg's adaptive rank-based analysis is also discussed. The chapter ends with discussion of the two-sample scale problem as well as a rank-based solution to the Behrens–Fisher problem. In Chapter 4, we discuss the rank-based procedures for regression models. We begin with simple linear regression and proceed to multiple regression. Besides fitting and diagnostic procedures to check the quality of fit, standard errors and tests of general linear hypotheses are discussed. Bootstrap methods and nonparametric regression models are also touched upon. This chapter closes with a presentation of Kendall's and Spearman's nonparametric correlation procedures. Many examples illustrate the computation of these procedures using R.

In Chapter 5, rank-based analysis and its computation for general fixed effects models are covered. Models discussed include one-way, two- and k-way designs, and analysis of covariance type designs, i.e., robust ANOVA and ANCOVA. The hypotheses tested by these functions are of Type III; that is, the tested effect is adjusted for all other effects. Multiple comparison procedures are an option for the one-way function. Besides rank-based analyses, we also cover the traditional Kruskal–Wallis one-way test and the ordered alternative problem including Jonckheere's test. The generalization of the Fligner–Killeen procedure to the k-sample scale problem is also covered.

Time-to-event analyses form the topic of Chapter 6. The chapter begins with a discussion of the Kaplan–Meier estimate and then proceeds to Cox's proportional hazards model and accelerated failure time models. The robust fitting methods for regression discussed in Chapter 4 are highly efficient procedures but they are sensitive to outliers in design space. In Chapter 7, high breakdown fits are presented for general regression models. These fits can attain up to 50% breakdown. Further, we discuss diagnostics which measure the difference between the highly efficient fits and the high breakdown fits of general linear models. We then consider these fits for nonlinear and time series models.

Rank-based inference for cluster correlated data is the topic of Chapter 8. The traditional Friedman's test is presented. Computational algorithms using R are presented for estimating the fixed effects and the variance components for these mixed effects models. Besides the rank-based fits discussed in Chapters 3–5, other types of R estimates are discussed. These include, for quite general covariance structure, GEERB estimates which are obtained by a robust iterated re-weighted least squares type of fit.

Besides Rfit, we have written the R package npsm which includes additional functions and datasets for methods presented in the first six chapters. Installing npsm and loading it at the start of each R session should allow the reader to reproduce all of these analyses. Topics in Chapters 7 and 8 require additional packages and details are provided in the text. The book itself was developed using Sweave (Leisch 2002) so the analyses have a high probability of being reproducible.

The first author would like to thank SDAC in general with particular thanks to Marian Fisher for her support of this effort, Tom Cook for thoughtful discussions, and Scott Diegel for general as well as technical assistance. In addition, he thanks KB Boomer, Mike Frey, and Jo Hardin for discussions on topics of statistics. The second author thanks Tom Hettmansperger and Simon Sheather for enlightening discussions on statistics throughout the years. For insightful discussions on rank-based procedures, he is indebted to many colleagues including Ash Abebe, Yusuf Bilgic, Magdalena Niewiadomska-Bugaj, Kim Crimin, Josh Naranjo, Jerry Sievers, Jeff Terpstra, and Tom Vidmar. We appreciate the efforts of John Kimmel of Chapman & Hall and, in general, the staff of Chapman & Hall for their help in the preparation of this book for publication. We are indebted to all who have helped make R a relatively easy to use but also very powerful computational language for statistics. We are grateful for our students' comments and suggestions when we developed parts of this material for lectures in robust nonparametric statistical procedures.

John Kloke
Joe McKean

1

Getting Started with R

This chapter serves as a primer for R. We invite the reader to start his or her R session and follow along with our discussion. We assume the reader is familiar with basic summary statistics and graphics taught in standard introductory statistics courses. We present a short tour of the langage; those interested in a more thorough introduction are referred to a monograph on R (e.g., Chambers 2008). Also, there are a number of manuals available at the Comprehensive R Archive Network (CRAN) (`http://cran.r-project.org/`). An excellent overview, written by developers of R, is Venables and Ripley (2002).

R provides a built-in documentation system. Using the help function i.e. `help(command)` or `?command` in your R session to bring up the help page (similar to a man page in traditional Unix) for the `command`. For example try: `help(help)` or `help(median)` or `help(rfit)`. Of course, Google is another excellent resource.

1.1   R Basics

Without going into a lot of detail, R has the capability of handling character (strings), logical (TRUE or FALSE), and of course numeric data types. To illustrate the use of R we multiply the system defined constant `pi` by 2.

```
> 2*pi
```

```
[1] 6.283185
```

We usually want to save the result for later calculation, so **assignment** is important. Assignment in R is usually carried out using either the `<-` operator or the `=` operator. As an example, the following code computes the area of a circle with radius 4/3 and assigns it to the variable `A`:

```
> r<-4/3
> A<-pi*r^2
> A
```

```
[1] 5.585054
```

1

In data analysis, suppose we have a set of numbers we wish to work with, as illustrated in the following code segment, we use the c operator to **combine** values into a vector. There are also functions rep for repeat and seq for sequence to create patterned data.

```
> x<-c(11,218,123,36,1001)
> y<-rep(1,5)
> z<-seq(1,5,by=1)
> x+y

[1]    12  219  124    37 1002

> y+z

[1] 2 3 4 5 6
```

The vector z could also be created with z<-1:5 or z<-c(1:3,4:5). Notice that R does vector arithmetic; that is, when given two lists of the same length it adds each of the elements. Adding a scalar to a list results in the scalar being added to each element of the list.

```
> z+10

[1] 11 12 13 14 15
```

One of the great things about R is that it uses logical naming conventions as illustrated in the following code segment.

```
> sum(y)

[1] 5

> mean(z)

[1] 3

> sd(z)

[1] 1.581139

> length(z)

[1] 5
```

Character data are embedded in quotation marks, either single or double quotes; for example, first<-'Fred' or last<-"Flintstone". The outcomes from the toss of a coin can be represented by

```
> coin<-c('H','T')
```

To simulate three tosses of a fair coin one can use the `sample` command

```
> sample(coin,3,replace=TRUE)
```

```
[1] "H" "T" "T"
```

The values `TRUE` and `FALSE` are reserved words and represent logical constants. The global variables `T` and `F` are defined as `TRUE` and `FALSE` respectively. When writing production code, one should use the reserved words.

1.1.1 Data Frames and Matrices

Data frames are a standard data object in R and are used to combine several variables of the same length, but not necessarily the same type, into a single unit. To combine `x` and `y` into a single data object we execute the following code.

```
> D<-data.frame(x,y)
> D
```

```
    x y
1   11 1
2  218 1
3  123 1
4   36 1
5 1001 1
```

To access one of the vectors the `$` operator may be used. For example to calculate the mean of `x` the following code may be executed.

```
> mean(D$x)
```

```
[1] 277.8
```

One may also use the column number or column name `D[,1]` or `D[,'x']` respectively. Omitting the first subscript means to use all rows. The `with` command as follows is another convenient alternative.

```
> with(D,mean(x))
```

```
[1] 277.8
```

As yet another alternative, many of the modeling functions in R have a `data=` options for which the data frame (or matrix) may be supplied. We utilize this option when we discuss regression modeling beginning in Chapter 4.

In data analysis, records often consist of mixed types of data. The following code illustrates combining the different types into one data frame.

```
> subjects<-c('Jim','Jack','Joe','Mary','Jean')
> sex<-c('M','M','M','F','F')
> score<-c(85,90,75,100,70)
> D2<-data.frame(subjects,sex,score)
> D2
```

```
  subjects sex score
1      Jim   M    85
2     Jack   M    90
3      Joe   M    75
4     Mary   F   100
5     Jean   F    70
```

Another variable can be added by using the $ operator for example
D2$letter<-c('B','A','C','A','C').

A set of vectors of the same type and size can be grouped into a matrix.

```
> X<-cbind(x,y,z)
> is.matrix(X)
```

```
[1] TRUE
```

```
> dim(X)
```

```
[1] 5 3
```

Note that R is case sensitive so that X is a different variable (or more generally, data object) than x.

1.2   Reading External Data

There are a number of ways to read data from an external file into R, for example scan or read.table. Though read.table and its variants (see help(read.table)) can read files from a local file system, in the following we illustrate the use of loading a file from the Internet. Using the command

egData<-read.csv('http://www.biostat.wisc.edu/~kloke/eg1.csv')

the contents of the dataset are now available in the current R session. To display the first several lines we may use the head command:

```
> head(egData)
```

```
  X       x1 x2          y
1 1 0.3407328  0 0.19320286
```

```
2 2 0.0620808  1   0.17166831
3 3 0.9105367  0   0.02707827
4 4 0.2687611  1  -0.78894410
5 5 0.2079045  0   9.39790066
6 6 0.9947691  1  -0.86209203
```

1.3 Generating Random Data

R has an abundance of methods for random number generation. The methods start with the letter r (for random) followed by an abbreviation for the name of the distribution. For example, to generate a pseudo-random list of data from normal (Gaussian) distribution, one would use the command rnorm. The following code segment generates a sample of size $n = 8$ of random variates from a standard normal distribution.

```
> z<-rnorm(8)
```

Often, in introductory statistics courses, to illustrate generation of data, the student is asked to toss a fair coin, say, 10 times and record the number of trials that resulted in heads. The following experiment simulates a class of 28 students each tossing a fair coin 10 times. Note that any text to right of the sharp (or pound) symbol # is completely ignored by R. i.e. represents a **comment**.

```
> n<-10
> CoinTosses<-rbinom(28,n,0.5)
> mean(CoinTosses) # should be close to 10*0.5 = 5

[1] 5.178571

> var(CoinTosses)  # should be close to 10*0.5*0.5 = 2.5

[1] 2.300265
```

In nonparametric statistics, often, a **contaminated normal** distribution is used to compare the robustness of two procedures to a violation of model assumptions. The contaminated normal is a mixture of two normal distributions, say $X \sim N(0,1)$ and $Y \sim N(0, \sigma_c^2)$. In this case X is a standard normal and both distributions have the same location parameter $\mu = 0$. Let ϵ denote the probability an observation is drawn from Y and $1 - \epsilon$ denote the probability an observation is drawn from X. The cumulative distribution function (cdf) of this model is given by

$$F(x) = (1 - \epsilon)\Phi(x) + \epsilon\Phi(x/\sigma_c) \tag{1.1}$$

where $\Phi(x)$ is the cdf of a standard normal distribution. In `npsm` we have included the function `rcn` which returns random deviates from this model. The `rcn` takes three arguments: `n` is the samples size (n), `eps` is the amount of contamination (ϵ), and `sigmac` is standard deviation of the contaminated part (σ_c). In the following code segment we obtain a sample of size $n = 1000$ from this model with $\epsilon = 0.1$ and $\sigma_c = 3$.

```
> d<-rcn(1000,0.1,3)
> mean(d)          # should be close to 0

[1] -0.02892658

> var(d)           # should be close to 0.9*1 + 0.1*9 = 1.8

[1] 2.124262
```

1.4 Graphics

R has some of the best graphics capabilities of any statistical software package; one can make high quality graphics with a few lines of R code. In this book we are using base graphics, but there are other graphical R packages available, for example, the R package `ggplot2` (Wickham 2009).

Continuing with the classroom coin toss example, we can examine the sampling distribution of the sample proportion. The following code segment generates the histogram of $\hat{p}$s displayed in Figure 1.1.

```
> phat<-CoinTosses/n
> hist(phat)
```

To examine the relationship between two variables we can use the `plot` command which, when applied to numeric objects, draws a scatterplot. As an illustration, we first generate a set of $n = 47$ datapoints from the linear model $y = 0.5x + e$ where $e \sim N(0, 0.1^2)$ and $x \sim U(0, 1)$.

```
> n<-47
> x<-runif(n)
> y<-0.5*x+rnorm(n,sd=0.1)
```

Next, using the the command `plot(x,y)` we create a simple scatterplot of `x` versus `y`. One could also use a `formula` as in `plot(y~x)`. Generally one will want to label the axes and add a title as the following code illustrates; the resulting scatterplot is presented in Figure 1.2.

```
> plot(x,y,xlab='Explanatory Variable',ylab='Response Variable',
+ main='An Example of a Scatterplot')
```

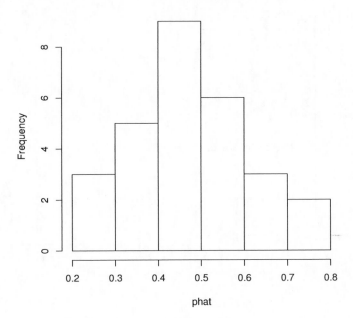

FIGURE 1.1
Histogram of 28 sample proportions; each estimating the proportion of heads in 10 tosses of a fair coin.

There are many options that can be set; for example, the plotting symbol, the size, and the color. Text and a legend may be added using the commands `text` and `legend`.

1.5   Repeating Tasks

Often in scientific computing a task is to be repeated a number of times. R offers a number of ways of replicating the same code a number of times making iterating straightforward. In this section we discuss the `apply`, `for`, and `tapply` functions.

The apply function will repeatedly apply a function to the rows or columns of a matrix. For example to calculate the mean of the columns of the matrix D previously defined we execute the following code:

```
> apply(D,2,mean)
```

An Example of a Scatterplot

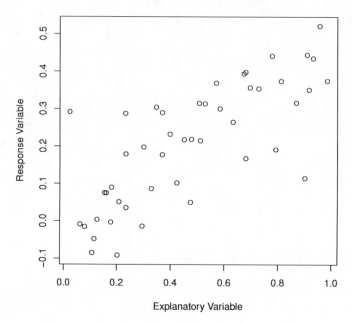

FIGURE 1.2
Example usage of the `plot` command.

```
      x       y
 277.8     1.0
```

To apply a function to the rows of a matrix the second argument would be set to 1. The `apply` function is discussed further in the next section in the context of Monte Carlo simulations.

A simple example to demonstrate the use of a `for` loop returns a vector of length n with the cumulative sum.

```
> n<-10
> result<-rep(1,n)
> for( i in 1:n ) result[i]<-sum(1:i)
```

Using `for` is discouraged in R; a loop generally results in much slower computational time than a vectorized function such as `apply`.

The function `tapply` is useful in obtaining summary statistics by cohort. For example, to calculate the mean score by sex from the D2 data we may use the `tapply` command.

```
> with(D2, tapply(score,sex,mean))
```

```
       F          M
85.00000 83.33333
```

A general purpose package for repeated tests for arrays, lists, matrices or data frames is `plyr` (Wickham 2011).

1.6   User Defined Functions

The syntax for creating an R function is relatively simple. A brief schematic for an R function is:

```
name_of_function <- function( 0 or more arguments ){

        ... body of function ...
}
```

where `name_of_function` contains the newly created function; the parentheses after `function` enclose the arguments of the function; and the braces { } enclose any number of R statements including function calls. A call to a user defined function is done in the expected way. E.g. `result<-name_of_function(data,arguments)`. Usually, the last line of the body contains a line of what is to be returned. We illustrate these concepts with the following example which computes the median and interquartile range of a sample contained in the data vector x. We named it `mSummary`.

```
mSummary <- function(x) {
    q1 <- quantile(x,.25)
    q3 <- quantile(x,.75)
    list(med=median(x),iqr=q3-q1)
}
```

These commands can be typed directly into an R session or copied and pasted from another file. Alternatively, the function may be sourced. If the function is in the file `mSummary.r` in the working directory, this can be accomplished by the R command `source("mSummary.r")`. If the file is in another directory (or folder) then the path to it must be included; the path may be relative or absolute. For example, if `mSummary.r` is in the directory `Myfunctions` which is a subdirectory of the current working directory then the command is `source("Myfunctions/mSummary.r")`. For a simple debugging run, we used the sample consisting of the first 13 positive integers.

```
> xsamp <- 1:13
> mSummary(xsamp)
```

```
$med
[1] 7

$iqr
75%
  6
```

Notice a list is returned with two elements: the median (`med`) and the IQR (`iqr`). A function need only be sourced once in an R session.

1.7   Monte Carlo Simulation

Simulation is a powerful tool in modern statistics. Inferences for rank-based procedures discussed in this book are based, generally, on the asymptotic distribution of the estimators. Simulation studies allow us to examine their performance for small samples. Specifically, simulation is used to examine the empirical level (or power) of rank-based tests of hypotheses or the empirical coverage of their confidence intervals. Comparisons of estimators are often based on their empirical relative efficiencies (the ratio of the mean squared error of the two estimators). Simulation is also used to examine the effect of violations of model assumptions on the validity of the rank-based inference. Another inference procedure used in this text is based on the bootstrap. This is a resampling technique, i.e., a Monte Carlo technique.

R is an excellent tool for simulation studies, because a simulation may be carried out with only a few lines of code. One way to run a simulation in R is to generate many samples from a distribution and use the apply function. For example,

```
> X<-matrix(rnorm(10*100),ncol=10)
```

generates a dataset with 100 rows and 10 columns. In the context of simulation, we think of the rows as distinct samples, each of size $n = 10$. To calculate the sample mean of each of the 100 samples, we use the apply function:

```
> xbar<-apply(X,1,mean)
```

The mean of each of the rows is calculated and the results are stored in the vector `xbar`. If we calculate the variance of the sample means we observe that it is similar to the theoretical result ($\sigma^2/n = 0.1$).

```
> var(xbar)
```

```
[1] 0.1143207
```

We can also do the same thing with the median

```
> xmed<-apply(X,1,median)
```

The relative efficiency is

```
> var(xbar)/var(xmed)
```

```
[1] 0.7146234
```

Exercise 1.9.4 asks the reader to compare the efficiency of these two estimators of location when the data are drawn from a t_3 distribution.

The **level** (α) of a statistical test is defined as the probability that the data support rejection of the null hypothesis when in fact the null hypothesis is true. The **power** of a statistical test is defined as the probability that the data support rejection of the null hypothesis when it is in fact false.

For our simple example, suppose we are interested in testing the null hypothesis that the true mean is 0. Using the 100 samples X, the following code obtains the empirical α-level of the nominal 5% t-test.

```
> myttest<-function(data) t.test(data)$p.value
> pval<-apply(X,1,myttest)
> mean(pval<0.05)
```

```
[1] 0.09
```

Exercise 1.9.11 asks the reader to approximate the power of the t-test under an alternative hypothesis.

1.8 R packages

The developers of R have made it fairly easy to extend R by creating a straightforward mechanism to create a **package**. A package may be developed for a small number of users or distributed worldwide. Two notable distribution sites are the Comprehensive R Archive Network (CRAN) and Bioconductor. The packages hosted at CRAN tend to be for general use while those hosted at Bioconductor are intended for analyzing high-throughput genomic data. At the time this book was to go to press the CRAN repository contained over 5500 packages developed by individual users.

We have written two such R packages related to nonparametrics: Rfit and npsm. Rfit (Kloke and McKean 2012) contains rank-based estimation and testing procedures for general linear models and is discussed extensively in Chapters 4 and 5. The package npsm includes many of the additional functions used in the book which are not already available in Rfit or base R. Most of the datasets used in this book are available in one of these packages. Both Rfit and npsm are available at CRAN. By loading npsm along with it's dependencies the reader will have the software tools and data nesessary to work through the first six chapters of the text. For later chapters, additional packages may be required and are available through `https://github.com/kloke/book`. New methods and features are being added to these packages and information will be available at that website. We anticipate any new code to be backward compatible to what is presented in this book. Two built-in R functions that help the user keep all their packages up-to-date are `new.packages` and `updated.packages`.

The `install.packages` command is a straightforward way to install a package from CRAN in an R session. For example to install the version of npsm on CRAN one could use the command

```
install.packages('npsm')
```

A pop-up window may appear asking the user to select a mirror. Once the mirror is selected (the user should use one that is close to him or her) R will then download npsm as well as any required packages, and then perform the installation. From then on the the package only needs to be loaded into R once per session using the function `library`. For example

```
library(npsm)
```

will load npsm and any packages on which it depends (e.g. `Rfit`).

1.9 Exercises

1.9.1. Use the commands `seq` and `rep` create the following lists.

    1. Even numbers less than 20

    2. Odd numbers between 101 and 203

    3. 1 3 1 3 1 3 1 3

    4. 1 1 1 1 3 3 3 3

1.9.2. Calculate the mean and variance of the following.

    1. First 100 integers.

    2. Random sample of 50 normal random variates with mean 30 and standard deviation 5.

1.9.3. Use the `sample` command to simulate a sequence of 10 tosses of a fair coin. Use 'H' to denote heads and 'T' to denote tails.

1.9.4. Using a t_3 distribution, approximate the relative efficiency of the sample median to the sample mean. Which estimator is more efficient for t_3 data?

1.9.5. Create a data frame D where the first column is named x and contains a vector of observed numeric values. Verify that the following commands all produce the same result.

1. `summary(D[1:nrow(D)],'x'])`
2. `summary(D[,'x'])`
3. `summary(D[!is.na(D$x),1])`
4. `summary(D[rep(TRUE,nrow(D)),1])`

1.9.6. What is the output for the command:
`rep(c(37,39,40,41,42),times=c(2,2,4,1,2))`?

1.9.7. A dotplot may be created with the command `stripchart` by using the option `method='stack'`. Create a dotplot of the data discussed in the previous exercise using the command `stripchart`.

1.9.8. A sunflower plot can be useful for visualizing the relationship between two numeric variables which are either discrete or have been rounded. Use the R function `sunflowerplot` to obtain a sunflower plot of the relationship between height and weight for the `baseball` data in `Rfit`.

1.9.9. A diagnostic test of clairvoyance is to declare a person clairvoyant if they get 8 or more tosses of a fair coin correct out of 10. Determine, either via simulation or directly, the specificity of the test. That is, in this case, determine the probability that a person who is guessing is correctly classified as non-clairvoyant.

1.9.10. Simulate the sampling distribution of the mean of 10 tosses of a fair die.

1.9.11. Approximate the power of a t-test of $H_0 : \mu = 0$ versus $H_A : \mu > 0$ when the true mean is $\mu = 0.5$. Assume a random sample of size $n = 25$ from a normal distribution with $\sigma = 1$. Assume $\alpha = 0.05$.

1.9.12. Use the commands `dnorm`, `seq`, and `lines` to create a plot of the pdf of a normal distribution with $\mu = 50$ and $\sigma^2 = 10$.

2

Basic Statistics

2.1 Introduction

In this chapter we present some basic nonparametric statistical procedures and show their computation in R. We begin with a brief example involving the distribution-free sign test. Then for the one-sample problem for continuous data, we present the signed-rank Wilcoxon nonparametric procedure and review the parametric t procedure. Next, we discuss inference based on bootstrapping (resampling). In the second part of the chapter, we turn our attention to discrete data. We discuss inference for the binomial probability models of the one- and two-sample problems, which we then generalize to the common goodness-of-fit χ^2-tests including the usual tests of homogeneity of distributions and independence for discrete random variables. We next present McNemar's test for significant change. We close the chapter with a brief discussion on robustness.

Our discussion focuses on the computation of these methods via R. More details of these nonparametric procedures can be found in the books by Hettmansperger and McKean (2011), Higgins (2003), Hollander and Wolfe (1999). A more theoretical discussion on the χ^2 goodness-of-fit tests can be found in Agresti (2002) or Hogg, McKean, and Craig (2013).

2.2 Sign Test

The sign test requires only the weakest assumptions of the data. For instance, in comparing two objects the sign test only uses the information that one object is better in some sense than the other.

As an example, suppose that we are comparing two brands of ice cream, say Brand A and Brand B. A blindfolded taster is given the ice creams in a randomized order with a washout period between tastes. His/her response is the preference of one ice cream over the other. For illustration, suppose that 12 tasters have been selected. Each taster is put through the blindfolded test. Suppose the results are such that Brand A is preferred by 10 of the tasters, Brand B by one of the tasters, and one taster has no preference. These data

15

present pretty convincing evidence in favor of Brand A. How likely is such a result due to chance if the null hypothesis is true, i.e., no preference in the brands? As our sign test statistic, let S denote the number of tasters that prefer Brand A to Brand B. Then for our data $S = 10$. The null hypothesis is that there is no preference in brands; that is, one brand is selected over the other with probability $1/2$. Under the null hypothesis, then, S has a binomial distribution with the probability of success of $1/2$ and, in this case, $n = 11$ as the number of trials. A two-sided p-value can be calculated as follows.

```
> 2*dbinom(10,11,1/2)
```

```
[1] 0.01074219
```

On the basis of this p-value we would reject the null hypothesis at the 5% level. If we make a one-sided conclusion, we would say Brand A is preferred over Brand B.

The sign test is an example of a distribution-free (nonparametric) test. In the example, suppose we can measure numerically the goodness of taste. The distribution of the sign test under the null hypothesis does not depend on the distribution of the measure; hence, the term distribution-free. The distribution of the sign test, though, is not distribution-free under alternative hypotheses.

2.3   Signed-Rank Wilcoxon

The sign test discussed in the last section is a nonparametric procedure for the one-sample or paired problem. Although it requires few assumptions, the power can be low, for example relative to the t-test at the normal distribution. In this section, we present the **signed-rank Wilcoxon** procedure which is a nonparametric procedure that has power nearly that of the t-test for normal distributions and it generally has power greater than that of the t-test for distributions with heavier tails than the normal distribution. More details for the Wilcoxon signed-rank test can be found in the references cited in Section 2.1. We discuss these two nonparametric procedures and the t-procedure for the one-sample location problem, showing their computation using R. For each procedure, we also discuss the R syntax for computing the associated estimate and confidence interval for the location effect.

We begin by defining a location model to set the stage for future discussions. Let $X_1, X_2, \ldots, X_n$ denote a random sample which follows the model

$$X_i = \theta + e_i, \tag{2.1}$$

where, to simplify discussion, we assume that the random errors, $e_1, \ldots, e_n$ are independent and identically distributed (iid) with a continuous probability density function $f(t)$ which is symmetric about 0. We call this model the

location model. Under the assumption of symmetry any location measure (parameter) of X_i, including the mean and median, is equal to θ. Suppose we are interested in testing the hypotheses

$$H_0 : \theta = 0 \text{ versus } H_A : \theta > 0. \tag{2.2}$$

The sign test of the last section is based on the test statistic

$$S = \sum_{i=1}^{n} \text{sign}(X_i), \tag{2.3}$$

where $\text{sign}(t) = -1, 0$, or 1 for $t < 0$, $t = 0$, or $t > 0$, respectively. Let

$$S^+ = \#_i\{X_i > 0\}. \tag{2.4}$$

Then $S = 2S^+ - n$. This assumes that none of the X_i's is equal to 0. In practice, generally observations with value 0 are omitted and the sample size is reduced accordingly. Note that under H_0, S^+ has a binomial distribution with n trials and probability of success $1/2$. Hence, critical values of the test are obtained from the binomial distribution. Since the null distribution of S does not depend on $f(t)$, we say that the sign test is **distribution-free**. Let s^+ denote the observed (realized) value of S^+ when the sample is drawn. Then the p-value of the sign test for the hypotheses (2.2) is $P_{H_0}(S^+ \geq s^+) = 1 - F_B(s^+ - 1; n, 0.5)$, where $F_B(t; n, p)$ denotes the cdf of a binomial distribution over n trials with probability of success p (pbinom is the R function which returns the cdf of a binomial distribution).

The traditional t-test of the hypotheses (2.2) is based on the sum of the observations.[1] The distribution of the statistic T depends on the population pdf $f(x)$. In particular, it is not distribution-free. The usual form of the test is the t-ratio

$$t = \frac{\overline{X}}{s/\sqrt{n}}, \tag{2.5}$$

where $\overline{X}$ and s are, respectively, the mean and standard deviation of the sample. If the population is normal then t has a Student t-distribution with $n-1$ degrees of freedom. Let t_0 be the observed value of t. Then the p-value of the t-test for the hypotheses (2.2) is $P_{H_0}(t \geq t_0) = 1 - F_T(t_0; n-1)$, where $F_T(t; \nu)$ denotes the cdf of the Student t-distribution with ν degrees of freedom (pt is the R function which returns the cdf of a t distribution). This is an exact p-value if the population is normal; otherwise it is an approximation.

The difference between the t-test and the sign test is that the t-test statistic is a function of the distances of the sample items from 0 in addition to their signs. The **signed-rank Wilcoxon** test statistic, however, uses only the ranks of these distances. Let $R|X_i|$ denote the rank of $|X_i|$ among $|X_1|, \ldots, |X_n|$,

---

[1] For comparison purposes, can be written as $T = \sum_{i=1}^{n} \text{sign}(X_i)|X_i|$.

from low to high. Then the signed-rank Wilcoxon test statistic is

$$W = \sum_{i=1}^{n} \text{sign}(X_i) R|X_i|. \tag{2.6}$$

Unlike the t-test statistic, W is distribution-free under H_0. Its distribution, though, cannot be obtained in closed-form. There are iterated algorithms for its computation which are implemented in R (`psignrank`, `qsignrank`, etc.). Usually the statistic computed is the sum of the ranks of the positive items, W^+, which is

$$W^+ = \sum_{X_i > 0} R|X_i| = \frac{1}{2}W + \frac{n(n+1)}{4}. \tag{2.7}$$

The R function `psignrank` computes the cdf of W^+. Let w^+ be the observed value of W^+. Then, for the hypotheses (2.2), the p-value of the signed-rank Wilcoxon test is $P_{H_0}(W^+ \geq w^+) = 1 - F_{W+}(w^+ - 1; n)$, where $F_{W+}(x; n)$ denotes the cdf of the signed-rank Wilcoxon distribution for a sample of size n.

2.3.1   Estimation and Confidence Intervals

Each of these three tests has an associated estimate and confidence interval for the location effect θ of Model (2.1). They are based on inversions[2] of the associated process. In this section we present the results and then show their computation in R. As in the last section, assume that we have modeled the sample $X_i, X_2, \ldots, X_n$ as the location model given in expression (2.1).

The confidence intervals discussed below, involve the order statistics of a sample. We denote the order statistics with the usual notation; that is, $X_{(1)}$ is the minimum of the sample, $X_{(2)}$ is the next smallest, $\ldots$, and $X_{(n)}$ is the maximum of the sample. Hence, $X_{(1)} < X_{(2)} < \cdots < X_{(n)}$. For example, if the sample results in $x_1 = 51, x_2 = 64, x_3 = 43$ then the ordered sample is given by $x_{(1)} = 43, x_{(2)} = 51, x_{(3)} = 64$.

The estimator of the location parameter θ associated with sign test is the sample median which we write as,

$$\hat{\theta} = \text{median}\{X_i, X_2, \ldots, X_n\}. \tag{2.8}$$

For $0 < \alpha < 1$, a corresponding confidence interval for θ of confidence $(1 - \alpha)100\%$ is given by $(X_{(c_1+1)}, X_{(n-c_1)})$, where $X_{(i)}$ denotes the ith order statistic of the sample and c_1 is the $\alpha/2$ quantile of the binomial distribution, i.e., $F_B(c_1; n, 0.5) = \alpha/2$; see Section 1.3 of Hettmansperger and McKean (2011) for details. This confidence interval is distribution-free and, hence, has exact confidence $(1 - \alpha)100\%$ for any random error distribution. Due to the discreteness of the binomial distribution, for each value of n, there is a limited number of values for α. Approximate interpolated confidence intervals for the median are presented in Section 1.10 of Hettmansperger and McKean (2011)

---

[2]See, for example, Chapter 1 of Hettmansperger and McKean (2011).

With regard to the t-test, the associated estimator of location is the sample mean $\overline{X}$. The usual confidence interval for θ is $(\overline{X} - t_{\alpha/2,n-1}[s/\sqrt{n}], \overline{X} + t_{\alpha/2,n-1}[s/\sqrt{n}])$, where $F_T(-t_{\alpha/2,n-1}; n-1) = \alpha/2$. This interval has the exact confidence of $(1-\alpha)100\%$ provided the population is normal. If the population is not normal then the confidence coefficient is approximately $(1 - \alpha)100\%$. Note the t-procedures are not distribution-free.

For the signed-rank Wilcoxon, the estimator of location is the Hodges–Lehmann estimator which is given by

$$\hat{\theta}_W = \text{med}_{i \le j} \left\{ \frac{X_i + X_j}{2} \right\}. \tag{2.9}$$

The pairwise averages $A_{ij} = (X_i + X_j)/2$, $i \le j$, are called the Walsh averages of the sample. Let $A_{(1)} < \cdots < A_{(n(n+1)/2)}$ denote the ordered Walsh averages. Then a $(1 - \alpha)100\%$ confidence interval for θ is

$$(A_{(c_2+1)}, A_{([n(n+1)/2]-c_2)}),$$

c_2 is the $\alpha/2$ quantile of the signed-rank Wilcoxon distribution. Provided the random error pdf is symmetric, this is a distribution-free confidence interval which has exact confidence $(1 - \alpha)100\%$. Note that the range of W^+ is the set $\{0, 1, \ldots n(n + 1)/2\}$ which is of order n^2. So for moderate sample sizes the signed-rank Wilcoxon does not have the discreteness problems that the inference based on the sign test has; meaning α is close to the desired level.

2.3.2 Computation in R

The signed-rank Wilcoxon and t procedures can be computed by the intrinsic R functions `wilcox.test` and `t.test`, respectively. Suppose x is the R vector containing the sample items. Then for the two-sided signed-rank Wilcoxon test of $H_0 : \theta = 0$, the call is

```
wilcox.test(x,conf.int=TRUE).
```

This returns the value of the test statistic W^+, the p-value of the test, the Hodges–Lehmann estimate of θ and the distribution-free 95% confidence interval for θ. The `t.test` function has similar syntax. The default hypothesis is two-sided. For the one-sided hypothesis $H_A : \theta > 0$, use `alternative="greater"` as an argument. If we are interested in testing the null hypothesis H_0 $\theta = 5$, for example, use `mu=5` as an argument. For, say, a 90% confidence interval use the argument `conf.level = .90`. For more information see the help page (`help(wilcox.test)`). Although the sign procedure does not have an intrinsic R function, it is simple to code such a function. One such R-function is given in Exercise 2.8.7.

Example 2.3.1 (Nursery School Intervention). This dataset is drawn from a study discussed by Siegel (1956). It involves eight pairs of identical twins who are of nursery school age. In the study, for each pair, one is randomly selected to attend nursery school while the other remains at home. At the end of the study period, all 16 children are given the same social awareness test. For each pair, the response of interest is the difference in the twins' scores, (Twin at School − Twin at Home). Let θ be the true median effect. As discussed in Remark 2.3.1, the random selection within a pair ensures that the response is symmetrically distributed under $H_0 : \theta = 0$. So the signed-rank Wilcoxon process is appropriate for this study. The following R session displays the results of the signed-rank Wilcoxon and the Student t-tests for one-sided tests of $H_0 : \theta = 0$ versus $H_A : \theta > 0$.

```
> school<-c(82,69,73,43,58,56,76,65)
> home<-c(63,42,74,37,51,43,80,62)
> response <- school - home
> wilcox.test(response,alternative="greater",conf.int=TRUE)

        Wilcoxon signed rank test

data:  response
V = 32, p-value = 0.02734
alternative hypothesis: true location is greater than 0
95 percent confidence interval:
   1 Inf
sample estimates:
(pseudo)median
         7.75

> t.test(response,alternative="greater",conf.int=TRUE)

        One Sample t-test

data:  response
t = 2.3791, df = 7, p-value = 0.02447
alternative hypothesis: true mean is greater than 0
95 percent confidence interval:
 1.781971        Inf
sample estimates:
mean of x
     8.75
```

Both procedures reject the null hypothesis at level 0.05. Note that the one-sided test option forces a one-sided confidence interval. To obtain a two-sided confidence interval use the two-sided option. ∎

Remark 2.3.1 (Randomly Paired Designs). The design used in the nursery school study is called a randomly paired design. For such a design, the experimental unit is a block of length two. In particular, in the nursery school study, the block was a set of identical twins. The factor of interest has two levels or there are two treatments. Within a block, the treatments are assigned at random, say, by a flip of a fair coin. Suppose H_0 is true; i.e., there is no treatment effect. If d is a response realization, then whether we observe d or $-d$ depends on whether the coin came up heads or tails. Hence, D and $-D$ have the same distribution; i.e., D is symmetrically distributed about 0. Thus the symmetry assumption for the signed-rank Wilcoxon test automatically holds. ∎

As a last example, we present the results of a small simulation study.

Example 2.3.2. Which of the two tests, the signed-rank Wilcoxon or the t-test, is the more powerful? The answer depends on the distribution of the random errors. Discussions of the asymptotic power of these two tests can be found in the references cited at the beginning of this chapter. In this example, however, we compare the powers of these two tests empirically for a particular situation. Consider the situation where the random errors of Model (2.1) have a t-distribution with 2 degrees of freedom. Note that it suffices to use a standardized distribution such as this because the tests and their associated estimators are equivariant to location and scale changes. We are interested in the two-sided test of $H_0 : \theta = 0$ versus $H_A : \theta \neq 0$ at level $\alpha = 0.05$. The R code below obtains 10,000 samples from this situation. For each sample, it records the p-values of the two tests. Then the empirical power of a test is the proportion of times its p-values is less than or equal to 0.05.

```
n = 30; df = 2; nsims = 10000; mu = .5; collwil = rep(0,nsims)
collt = rep(0,nsims)
for(i in 1:nsims){
    x = rt(n,df) + mu
    wil = wilcox.test(x)
    collwil[i] = wil$p.value
    ttest = t.test(x)
    collt[i] = ttest$p.value
}
powwil = rep(0,nsims); powwil[collwil <= .05] = 1
powerwil = sum(powwil)/nsims
powt = rep(0,nsims); powt[collt <= .05] = 1
powert = sum(powt)/nsims
```

We ran this code for the three situations: $\theta = 0$ (null situation) and the two alternative situations with $\theta = 0.5$ and $\theta = 1$. The empirical powers of the tests are:

Test	$\theta = 0$	$\theta = 0.5$	$\theta = 1$
Wilcoxon	0.0503	0.4647	0.9203
t	0.0307	0.2919	0.6947

The empirical α level of the signed-rank Wilcoxon test is close to the nominal value of 0.05, which is not surprising because it is a distribution-free test. On the other hand, the t-test is somewhat conservative. In terms of power, the signed-rank Wilcoxon test is much more powerful than the t-test. So in this situation, the signed-rank Wilcoxon is the preferred test. ∎

2.4　Bootstrap

As computers have become more powerful, the bootstrap, as well as resampling procedures in general, has gained widespread use. The bootstrap is a general tool that is used to measure the error in an estimate or the significance of a test of hypothesis. In this book we demonstrate the bootstrap for a variety of problems, though we still only scratch the surface; the reader interested in a thorough treatment is referred to Efron and Tibshirani (1993) or Davison and Hinkley (1997). In this section we illustrate estimation of confidence intervals and p-values for the one-sample and paired location problems.

To fix ideas, recall a histogram of the sample is often used to provide context of the distribution of the random variable (e.g., location, variability, shape). One way to think of the bootstrap is that it is a procedure to provide some context for the the sampling distribution of a statistic. A bootstrap sample is simply a sample from the original sample taken with replacement. The idea is that if the sample is representative of the population, or more concretely, the histogram of the sample resembles the pdf of the random variable, then sampling from the sample is representative of sampling from the population. Doing so repeatedly will yield an estimate of the sampling distribution of the statistic.

R offers a number of capabilities for implementing the bootstrap. We begin with an example which illustrates the bootstrap computed by the R function `sample`. Using `sample` is useful for illustration; however, in practice one will likely want to implement one of R's internal functions and so the library `boot` (Canty and Ripley 2013) is also discussed.

Example 2.4.1. To illustrate the use of the bootstrap, first generate a sample of size 25 from a normal distribution with mean 30 and standard deviation 5.

```
> x<-rnorm(25,30,5)
```

In the following code segment we obtain 1000 bootstrap samples and for each sample we calculate the sample mean. The resulting vector `xbar` contains the 1000 sample means. Figure 2.1 contains a histogram of the 1000 estimates. We have also included a plot of the true pdf of the sampling distribution of $\bar{X}$; i.e. a $N(30, 5^2/25)$.

```
> B<-1000 # number of bootstrap samples to obtain
> xbar<-rep(0,B)
```

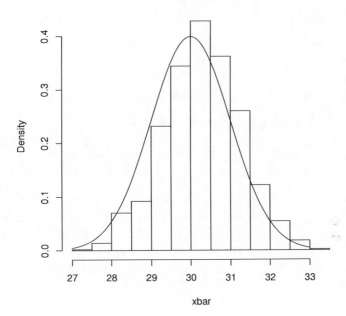

FIGURE 2.1

Histogram of 1000 bootstrap estimates of the sample mean based on a sample of size $n = 25$ from a $N(30, 5^2)$ distribution. The pdf of a $N(30, 1)$ is overlaid.

```
> for( i in 1:B ) {
+ xbs<-sample(x,length(x),replace=TRUE)
+ xbar[i]<-mean(xbs)
+ }
```

The standard deviation of the bootstrap sampling distribution may serve as an estimate of the standard error of the estimate.

```
> se.xbar<-sd(xbar)
> se.xbar
```

```
[1] 0.9568816
```

The estimated standard error may then be used for inference. For example, as we know the distribution of the sample mean is normally distributed, we can calculate an approximate 95% confidence interval using t-critical values as follows. We have included the usual t-interval for comparison.

```
> tcv<-qt(0.975,length(x)-1)
> mean(x)+c(-1,1)*tcv*se.xbar
```

```
[1] 28.12227 31.87773

> mean(x)+c(-1,1)*tcv*sd(x)/sqrt(length(x))

[1] 29.89236 30.10764
```

∎

2.4.1   Percentile Bootstrap Confidence Intervals

In Example 2.4.1 we presented a simple confidence interval based on a bootstrap estimate of standard error. Such an estimate requires assumptions on the sampling distribution of the estimate; for example, that the sampling distribution is symmetric and that the use of t-critical values is appropriate. In this section we present an alternative, the percentile bootstrap confidence interval, which is free of such assumptions. Let $\hat{\theta}$ be any location estimator.

Let $x = [x_1, \ldots, x_n]^T$ denote a vector of observations observed from the distribution F. Let $\hat{\theta}$ denote the estimate of θ based on this sample. Define the empirical cumulative distribution function of the sample by

$$\hat{F}_n(t) = \frac{1}{n} \sum_{i=1}^{n} I(x_i \leq t). \tag{2.10}$$

Then a **bootstrap sample** is a sample taken with replacement from $\hat{F}_n$; i.e. $x_1^*, \ldots x_n^*$ are iid $\hat{F}_n$. Denote this sample by $x^* = [x_1^*, \ldots, x_n^*]^T$. Let $\hat{\theta} = T(x)$ be the estimate based on the original sample. Similarly $\hat{\theta}^* = T(x^*)$ is the estimate based on the bootstrap sample. The bootstrap process is repeated a large number of times, say B, from which we obtain $\hat{\theta}_1^*, \ldots, \hat{\theta}_B^*$. Since the empirical distribution of the bootstrap estimates approximates the sampling distribution of $\hat{\theta}$ we may use it to obtain an estimate of certainty in our estimate $\hat{\theta}$. To obtain our confidence interval, we order the estimates $\hat{\theta}_{(1)}^* \leq \hat{\theta}_{(2)}^* \leq, \ldots, \leq \hat{\theta}_{(B)}^*$. Let $m = [\alpha/2 * B]$ then $(\hat{\theta}_{(m)}^*, \hat{\theta}_{(B-m)}^*)$ is an approximate $(1 - \alpha) * 100\%$ confidence interval for θ. That is, the end points of the **percentile bootstrap confidence interval** are the $\alpha/2$ and $1 - \alpha/2$ percentiles of the empirical distribution of the $\hat{\theta}_i^*$'s.

Returning again to our example, let $T(x) = \frac{1}{n} \sum_{i=1}^{n} x_i$. The following code segment obtains a 95% bootstrap percentile confidence interval.

```
> quantile(xbar,probs=c(0.025,0.975),type=1)

    2.5%     97.5%
28.30090 32.14894

> m<-0.025*1000
> sort(xbar)[c(m,B-m)]
```

[1] 28.30090 32.14894

Next we illustrate the use of the `boot` library to arrive at the same result.

```
> bsxbar<-boot(x,function(x,indices) mean(x[indices]), B)
> boot.ci(bsxbar)
```

```
BOOTSTRAP CONFIDENCE INTERVAL CALCULATIONS
Based on 1000 bootstrap replicates

CALL :
boot.ci(boot.out = bsxbar)

Intervals :
Level      Normal                 Basic
95%    (29.89, 30.11 )     (29.89, 30.11 )


Level      Percentile              BCa
95%    (29.89, 30.11 )     (29.89, 30.12 )
Calculations and Intervals on Original Scale
```

```
> quantile(bsxbar$t,probs=c(0.025,0.975),type=1)

    2.5%      97.5%
29.88688 30.11383
```

2.4.2   Bootstrap Tests of Hypotheses

In bootstrap testing, the resampling is conducted under conditions that ensure the null hypothesis, H_0, is true. This allows the formulation of a bootstrap p-value. In this section we illustrate the use of the bootstrap testing procedure by applying it to the paired and one-sample problems discussed in Sections 2.2–2.3.

The bootstrap testing procedure for the paired problem is as follows. First sample with replacement from the set of pairs; then treatment is applied at random to the pair. Notice this preserves the correlation of the paired design. If $d_1, \ldots, d_n$ denote the difference based on the original sample data, then in the bootstrap sample, if the ith pair is selected d_i and $-d_i$ each have probability $\frac{1}{2}$ of being in the bootstrap sample; hence the null hypothesis is true. Let $T_1^*, \ldots, T_B^*$ be the test statistics based on the B bootstrap samples. These form an estimate of the null distribution of the test statistic T. The bootstrap p-value is then calculated as

$$p\text{-value} = \frac{\#\{T_i^* \geq T\}}{B}.$$

Example 2.4.2 (Nursery School Intervention Continued). There is more than one way to to implement the bootstrap testing procedure for the paired problem, the following is one which utilizes the set of differences.

```
> d<-school-home
> dpm<-c(d,-d)
```

Then **dpm** contains all the $2n$ possible differences. Obtaining bootstrap samples from this vector ensures the null hypothesis is true. In the following we first obtain $B = 5000$ bootstrap samples and store them in the vector dbs.

```
> n<-length(d)
> B<-5000
> dbs<-matrix(sample(dpm,n*B,replace=TRUE),ncol=n)
```

Next we will use the **apply** function to obtain the Wilcoxon test statistic for each bootstrap sample. First we define a function which will return the value of the test statistic.

```
> wilcox.teststat<-function(x) wilcox.test(x)$statistic
> bs.teststat<-apply(dbs,1,wilcox.teststat)
> mean(bs.teststat>=wilcox.teststat(d))
```

[1] 0.0238

Hence, the p-value $= 0.0238$ and is significant at the 5% level. ■

For the second problem, consider the one-sample location problem where the goal is to test the hypothesis

$$H_0 : \theta = \theta_0 \text{ versus } \theta > \theta_0.$$

Let $x_1, \ldots, x_n$ be a random sample from a population with location θ. Let $\hat{\theta} = T(\boldsymbol{x})$ be an estimate of θ.

To ensure the null hypothesis is true, that we are sampling from a distribution with location θ_0, we take our bootstrap samples from

$$x_1 - \hat{\theta} + \theta_0, \ldots, x_n - \hat{\theta} + \theta_0. \tag{2.11}$$

Denote the bootstrap estimates as $\hat{\theta}_1^*, \ldots, \hat{\theta}_B^*$. Then the bootstrap p-value is given by

$$\frac{\#\{\hat{\theta}_i^* \geq \hat{\theta}\}}{B}.$$

We illustrate this bootstrap test with the following example.

Example 2.4.3 (Bootstrap test for sample mean). In the following code segment we first take a sample from a $N(1.5, 1)$ distribution. Then we illustrate a test of the hypothesis

$$H_0 : \theta = 1 \text{ versus } H_A : \theta > 1.$$

The sample size is $n = 25$.

```
> x<-rnorm(25,1.5,1)
> thetahat<-mean(x)
> x0<-x-thetahat+1 #theta0 is 1
> mean(x0) # notice H0 is true
```

```
[1] 1
```

```
> B<-5000
> xbar<-rep(0,B)
> for( i in 1:B ) {
+ xbs<-sample(x0,length(x),replace=TRUE)
+ xbar[i]<-mean(xbs)
+ }
> mean(xbar>=thetahat)
```

```
[1] 0.02
```

In this case the p-value $= 0.02$ is significant at the 5% level. ∎

2.5 Robustness*

In this section, we briefly discuss the robustness properties of the three estimators discussed so far in this chapter, namely, the mean, the median, and the Hodges–Lehmann estimate. Three of the main concepts in robustness are efficiency, influence, and breakdown. In Chapter 3, we touch on efficiency, while in this section we briefly explore the other two concepts for the three estimators.

The finite sample version of the influence function of an estimator is its sensitivity curve. It measures the change in an estimator when an outlier is added to the sample. More formally, let the vector $x_n = (x_1, x_2, \ldots, x_n)^T$ denote a sample of size n. Let $\hat{\theta}_n = \hat{\theta}_n(x_n)$ denote an estimator. Suppose we add a value x to the sample to form the new sample $x_{n+1} = (x_1, x_2, \ldots, x_n, x)^T$ of size $n + 1$. Then the **sensitivity curve** for the estimator is defined by

$$S(x; \hat{\theta}) = \frac{\hat{\theta}_{n+1} - \hat{\theta}_n}{1/(n+1)}. \tag{2.12}$$

The value $S(x; \hat{\theta})$ measures the rate of change of the estimator at the outlier x.

As an illustration consider the sample

$$\{1.85, 2.35, -3.85, -5.25, -0.15, 2.15, 0.15, -0.25, -0.55, 2.65\}.$$

The sample mean of this dataset is -0.09 while the median and Hodges–Lehmann estimates are both 0.0. The top panel of Figure 2.2 shows the sensitivity curves of the three estimators, the mean, median, and Hodges–Lehmann

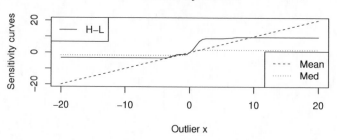

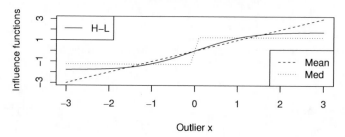

FIGURE 2.2

The top panel shows the sensitivity curves for the mean, median, and Hodges–Lehmann estimators for the sample given in the text. The bottom panel displays the influence function of the three estimators.

for this sample when x is in the interval $(-20, 20)$. Note that the sensitivity curve for the mean is unbounded; i.e., as the outlier becomes large the rate in change of the mean becomes large; i.e., the curve is unbounded. On the other hand, the median and the Hodges–Lehmann estimators change slightly as x changes sign, but their changes soon become constant no matter how large $|x|$ is. The sensitivity curves for the median and Hodges–Lehmann estimates are bounded.

While intuitive, a sensitivity curve depends on the sample items. Its theoretical analog is the influence function which measures rate of change of the functional of the estimator. at the probability distribution, $F(t)$, of the random errors of the location model. We say an estimator is **robust** if its influence function is bounded. Down to a constant of proportionality and center, the influence functions of the mean, median, and Hodges–Lehmann estimators at a point x are respectively x, $\text{sign}(x)$, and $F(x) - 0.5$. Hence, the median and the Hodges–Lehmann estimators are robust, while the mean is not. The influence functions of the three estimators are displayed in the lower panel of Figure 2.2 for a normal probability model. For the median and Hodges–Lehmann estimators, they are smooth versions of their respective sensitivity curves.

To briefly define the breakdown point of an estimator, consider again a sample $\boldsymbol{x}_n = (x_1, x_2, \ldots, x_n)^T$ from a location model with parameter θ. Let $\hat{\theta} = \hat{\theta}(\boldsymbol{x}_n)$ be an estimator of θ. Suppose we contaminate m points in the sample, so that the sample becomes

$$\boldsymbol{x}_n^* = (x_1^*, \ldots, x_m^*, x_{m+1}, \ldots, x_n)^T,$$

where $x_1^*, \ldots, x_m^*$ are the contaminated points. Think of the contaminated points as very large (nearly ∞) in absolute value. The smallest value of m so that the value of the estimator $\hat{\theta}(\boldsymbol{x}_n^*)$ becomes meaningless is the breaking point of the estimator and the ratio m/n is called the finite sample breakdown point of $\hat{\theta}$. If this ratio converges to a finite value, we call this value the **breakdown point** of the estimator. Notice for the sample mean that one point of contamination suffices to make the mean meaningless (as $x_1^* \to \infty$, $\bar{x} \to \infty$). Hence, the breakdown point of the mean is 0. On the other hand, the sample median can tolerate almost half of the data being contaminated. In the limit, its ratio converges to 0.50. So we say that median has 50% breakdown. The Hodges–Lehmann estimate has breakdown point 0.29; see, for instance, Chapter 1 of Hettmansperger and McKean (2011).

In summary, the sample median and the Hodges–Lehmann estimator are robust, with positive breakdown points. The mean is not robust and has breakdown point 0. Of the two, the sample median and the Hodges–Lehmann estimator, because of its higher breakdown, it would seem that the median is preferred. This, however, ignores efficiency between estimators which is discussed in Chapter 3. Efficiency generally reverses this preference.

2.6 One- and Two-Sample Proportion Problems

For this and the next section, we focus on discrete variables. Recall that X is a discrete random variable if its range consists of categories. In this section, we consider discrete random variables whose ranges consist of two categories which we generally label as failure (0) and success (1). Let X denote such a random variable. Let p denote the probability of success. Then we say that X has a Bernoulli distribution with the probability model

x	0	1
$P(X = x)$	$1 - p$	p

It is easy to show that the mean of X is p and that the variance of X is $p(1 - p)$.

2.6.1   One-Sample Problems

Statistical problems consist of estimating p, forming confidence intervals for it, and testing hypotheses of the form

$$H_0 : p = p_0 \text{ versus } H_A : p \neq p_0, \tag{2.13}$$

where p_0 is specified. One sided hypotheses can be similarly formulated.

Let $X_1, \ldots, X_n$ be a random sample on X. Let S be the total number of successes in the sample of size n. Then S has a binomial distribution with the distribution

$$P(S = j) = \binom{n}{j} p^j (1 - p)^{n-j}, \quad j = 0, 1, \ldots, n. \tag{2.14}$$

The estimate of p is the sample proportion of successes; i.e.,

$$\hat{p} = \frac{S}{n}. \tag{2.15}$$

Based on the asymptotic normality of S, an approximate $(1 - \alpha)100\%$ confidence interval for p is

$$(\hat{p} - z_{\alpha/2}\sqrt{\hat{p}(1 - \hat{p})/n}, \hat{p} + z_{\alpha/2}\sqrt{\hat{p}(1 - \hat{p})/n}). \tag{2.16}$$

Example 2.6.1 (Squeaky Hip Replacements). As a numerical example, Devore (2012), page 284, reports on a study of 143 subjects who have obtained ceramic hip replacements. Ten of the subjects in the study reported that their hip replacements squeaked. Consider patients who receive such a ceramic hip replacement and let p denote the true proportion of those whose replacement hips develop a squeak. Based on the data, we next compute[3] the estimate of p and a confidence interval for it.

```
> phat<-10/143
> zcv<-qnorm(0.975)
> phat+c(-1,1)*zcv*sqrt(phat*(1-phat)/143)

[1] 0.02813069 0.11172945
```

Hence, we estimate between roughly 3 and 11% of patients who receive ceramic hip replacements such as the ones in the study will report squeaky replacements. ■

Asymptotic tests of hypotheses involving proportions, such as (2.13), are often used. For hypotheses (2.13), the usual test is to reject H_0 in favor of H_A, if $|z|$ is large, where

$$z = \frac{\hat{p} - p_0}{\sqrt{p_0(1 - p_0)/n}} \tag{2.17}$$

---

[3]The base R function `prop.test` provides a confidence interval which is computed by inverting the score test.

Note that z has an asymptotic $N(0,1)$ distribution under H_0, so an equivalent test statistic is based on $\chi^2 = z^2$. The p-value for a two-sided test is p-value $= P[\chi^2(1) > (\text{Observed } \chi^2)]$. This χ^2-formulation is the test and p-value computed by the R function `prop.test` with `correct=FALSE` indicating that a continuity correction not be applied. The two-sided hypothesis is the default, but one-sided hypotheses can be tested by specifying the `alternative` argument. The null value of p is set by the argument p.

Example 2.6.2 (Left-Handed Professional Ball Players). As an example of this test, consider testing whether the proportion of left-handed professional baseball players is the same as the proportion of left-handed people in the general population, which is about 0.15. For our sample we use the dataset `baseball` that consists of observations on 59 professional baseball players, including throwing hand ('L' or 'R'). The following R segment computes the test:

```
> ind<-with(baseball,throw=='L')
> prop.test(sum(ind),length(ind),p=0.15,correct=FALSE)


        1-sample proportions test without continuity correction

data:  sum(ind) out of length(ind), null probability 0.15
X-squared = 5.0279, df = 1, p-value = 0.02494
alternative hypothesis: true p is not equal to 0.15
95 percent confidence interval:
 0.1605598 0.3779614
sample estimates:
        p
0.2542373
```

Because the p-value of the test is 0.02494, H_0 is rejected at the 5% level. ∎

The above inference is based on the asymptotic distribution of S, the number of successes in the sample. This statistic, though, has a binomial distribution, (2.14), and inference can be formulated based on it. This includes finite sample tests and confidence intervals. For a given level α, though, these confidence intervals are conservative; that is, their true confidence level is at least (see Section 4.3 of Hogg et al. (2013).) $1 - \alpha$. These tests and confidence intervals are computed by the R function `binom.test`. We illustrate its computation for the baseball example.

```
> binom.test(sum(ind),59,p=.15)

        Exact binomial test

data:  sum(ind) and 59
number of successes = 15, number of trials = 59, p-value = 0.04192
alternative hypothesis: true probability of success is not equal to 0.15
```

```
95 percent confidence interval:
 0.1498208 0.3844241
sample estimates:
probability of success
          0.2542373
```

Note that the confidence interval traps $p = 0.15$, even though the two-sided test rejects H_0. This example illustrates the conservativeness of the finite sample confidence interval.

2.6.2   Two-Sample Problems

Consider two Bernoulli random variables X and Y with respective probabilities of success p_1 and p_2. The parameter of interest is the difference in proportions $p_1 - p_2$. Inference concerns estimates of this difference, confidence intervals for it, and tests of hypotheses of the form

$$H_0 : \ p_1 = p_2 \text{ versus } H_A : \ p_1 \neq p_2. \tag{2.18}$$

Let $X_1, \ldots, X_{n_1}$ and $Y_1, \ldots, Y_{n_2}$ be random samples on X and Y, respectively. Assume that the samples are independent of one another. Section 2.7.4 discusses the paired (dependent) case. The estimate of the difference in proportions is the difference in sample proportions, i.e., $\hat{p}_1 - \hat{p}_2$.

It follows that the estimator $\hat{p}_1 - \hat{p}_2$ has an asymptotic normal distribution. Based on this, a $(1 - \alpha)100\%$ asymptotic confidence interval for $p_1 - p_2$ is

$$\hat{p}_1 - \hat{p}_2 \pm z_{\alpha/2} \sqrt{\frac{\hat{p}_1(1 - \hat{p}_1)}{n_1} + \frac{\hat{p}_2(1 - \hat{p}_2)}{n_2}}. \tag{2.19}$$

For the hypothesis (2.18), there are two test statistics which are used in practice. The Wald-type test is the standardization of $\hat{p}_1 - \hat{p}_2$ based on its standard error, (the square-root term in expression (2.19)). The more commonly used test statistic is the scores test which standardizes under H_0. Under H_0 the population proportions are the same; hence the following average

$$\hat{p} = \frac{n_1 \hat{p}_1 + n_2 \hat{p}_2}{n_1 + n_2} \tag{2.20}$$

is an estimate of the common proportion. The scores test statistic is given by

$$z = \frac{\hat{p}_1 - \hat{p}_2}{\sqrt{\hat{p}(1 - \hat{p})}\sqrt{\frac{1}{n_1} + \frac{1}{n_2}}}. \tag{2.21}$$

This test statistic is compared with z-critical values. As in the one-sample problem, the χ^2-formulation, $\chi^2 = z^2$, is often used. We illustrate the R computation of this analysis with the next example.

Basic Statistics

Example 2.6.3 (Polio Vaccine). Rasmussen (1992), page 355, discusses one of the original clinical studies for the efficacy of the Salk polio vaccine which took place in 1954. The effectiveness of the vaccine was not known and there were fears that it could even cause polio since the vaccine contained live virus. Children with parental written consent were randomly divided into two groups. Children in the treatment group (1) were injected with the vaccine while those in the control or placebo group (2) were injected with a biologically inert solution. Let p_1 and p_2 denote the true proportions of children who get polio in the treatment and control groups, respectively. The hypothesis of interest is the two-sided hypothesis (2.18). The following data are taken from Rasmussen (1992):

Group	No. Children	No. Polio Cases	Sample Proportion
Treatment	200,745	57	0.000284
Control	201,229	199	0.000706

The R function for the analysis is the same function used in the one-sample proportion problem, namely, `prop.test`. The first and second arguments are respectively the vectors `c(S1,S2)` and `c(n1,n2)`, where S1 and S2 are the number of successes for the two samples. The default hypotheses are the two-sided hypotheses (2.18). The following R segment provides the analysis for the polio vaccine data.

```
> prop.test(c(57,199),c(200745,201229),correct=FALSE)

        2-sample test for equality of proportions without
        continuity correction

data:  c(57, 199) out of c(200745, 201229)
X-squared = 78.4741, df = 1, p-value < 2.2e-16
alternative hypothesis: two.sided
95 percent confidence interval:
 -0.0008608391 -0.0005491224
sample estimates:
      prop 1        prop 2
0.0002839423 0.0009889231
```

The χ^2 test statistic has the value 77.3704 with a p-value that is zero to 15 places; hence, the null hypothesis would certainly be rejected. The direction indicates that the vaccine has been effective. ∎

2.7 χ^2 Tests

In Section 2.6.1, we discussed inference for Bernoulli (binomial) random variables; i.e., discrete random variables with a range consisting of two categories. In this section, we extend this discussion to discrete random variables whose range consists of a general number of categories. Recall that the tests of Section 2.6.1 could be formulated in terms of χ^2-tests. We extend these χ^2 goodness-of-fit tests for the situations of this section. Technical details may be found in Agresti (2002) or Hogg et al. (2013). Consider a hypothesis (null) concerning the categories. Then the χ^2-test statistic is essentially the sum over the categories of the squared and standardized differences between the observed and expected frequencies, where the expected frequencies are formulated under the assumption that the null hypothesis is true. In general, under the null hypothesis, this test statistic has an asymptotic χ^2-distribution with degrees of freedom equal to the number of free categories (cells) minus the number of parameters, if any, that need to be estimated to form the expected frequencies. As we note later, at times the exact null distribution can be used instead of the asymptotic distribution. For now, we present three general applications and their computation using R.

2.7.1 Goodness-of-Fit Tests for a Single Discrete Random Variable

Consider a discrete random variable X with range (categories) $\{1, 2, \ldots, c\}$. Let $p(j) = P[X = j]$ denote the the probability mass function (pmf) of the distribution of X. Suppose the hypotheses of interest are:

$$H_0 : p(j) = p_0(j), j = 1, \ldots, c \text{ versus } H_A : p(j) \neq p_0(j), \text{ for some } j. \quad (2.22)$$

Suppose $X_1, \ldots, X_n$ is a random sample on X. Let $O_j = \#\{X_i = j\}$. The statistics $O_1, \ldots O_c$ are called the observed frequencies of the categories of X. The observed frequencies are constrained as $\sum_{j=1}^{c} O_j = n$; so, there are essentially $c - 1$ free cells. The expected frequencies of these categories under H_0 are given by $E_j = E_{H_0}[O_j]$, where E_{H_0} denotes expectation under the null hypothesis. There are two cases.

In the first case, the null distribution probabilities, $p_0(j)$, are completely specified. In this case, $E_j = np_0(j)$ and the test statistic is given by

$$\chi^2 = \sum_{j=1}^{c} \frac{(O_j - E_j)^2}{E_j}. \quad (2.23)$$

The hypothesis H_0 is rejected in favor of H_A for large values of χ^2. Note that the vector of observed frequencies, $(O_1, \ldots, O_c)^T$ has a multinomial distribution, so the exact distribution of χ^2 can be obtained. It is also asymptotically equivalent to the likelihood ratio test statistic,[4] and, hence, has an asymptotically χ^2-distribution with $c-1$ degrees of freedom under H_0. In practice, this asymptotic result is generally used. Let χ_0^2 be the realized value of the statistic χ^2 when the sample is drawn. Then the p-value of the goodness-of-fit test is $1 - F_{\chi^2}(\chi_0^2; c-1)$, where $F_{\chi^2}(\cdot; c-1)$ denotes the cdf of a χ^2-distribution with $c-1$ degrees of freedom.

The R function `chisq.test` computes the test statistic (2.23). The input consists of the vectors of observed frequencies and the pmf $(p_0(1), \ldots, p_0(c))^T$. The uniform distribution $(p(j) \equiv 1/c)$ is the default null distribution. The output includes the value of the χ^2-test statistic and the p-value. The return list includes values for the observed frequencies (`observed`), the expected frequencies (`expected`), and the residuals (`residuals`). These residuals are $(O_j - E_j)/\sqrt{E_j}$, $j = 1, \ldots, c$ and are often called the Pearson residuals. The squares of the residuals are the categories' contributions to the test statistic and offer valuable post-test information on which categories had large discrepancies from those expected under H_0.

Here is a simple example. Suppose we roll a die $n = 370$ times and we observe the frequencies $(58, 55, 62, 68, 66, 61)^T$. Suppose we are interested in testing to see if the die is fair; i.e., $p(j) \equiv 1/6$. Computation in R yields

```
> x <- c(58,55,62,68,66,61)
> chifit <- chisq.test(x)
> chifit

        Chi-squared test for given probabilities

data:  x
X-squared = 1.9027, df = 5, p-value = 0.8624

> round(chifit$expected,digits=4)

[1]  61.6667 61.6667 61.6667 61.6667 61.6667 61.6667

> round((chifit$residuals)^2,digits=4)

[1]  0.2180 0.7207 0.0018 0.6505 0.3045 0.0072
```

Thus there is no evidence to support the die being unfair.

In the second case for the goodness-of-fit tests, only the form of the null pmf is known. Unknown parameters must be estimated.[5] Then the expected values are the estimates of E_j based on the estimated pmf. The degrees of freedom, though, decrease by the number of parameters that are estimated.[6] The following example illustrates this case.

---

[4]See, for example, Exercise 6.5.8 of Hogg et al. (2013).

[5]For this situation, generally we estimate the unknown parameters of the pmf by their maximum likelihood estimators. See Hogg et al. (2013).

[6]See Section 4.7 of Hogg et al. (2013).

Example 2.7.1 (Birth Rate of Males to Swedish Ministers). This data is discussed on page 266 of Daniel (1978). It concerns the number of males in the first seven children for $n = 1334$ Swedish ministers of religion. The data are

No. of Males	0	1	2	3	4	5	6	7
No. of Ministers	6	57	206	362	365	256	69	13

For example, 206 of these ministers had 2 sons in their first 7 children. The null hypothesis is that the number of sons is binomial with probability of success p, where success is a son. The maximum likelihood estimator of p is the number of successes over the total number of trials which is

$$\hat{p} = \frac{\sum_{j=0}^{7} j \times O_j}{7 \times 1334} = 0.5140.$$

The expected frequencies are computed as

$$E_j = n \binom{7}{j} \hat{p}^j (1 - \hat{p})^{7-j}.$$

The values of the pmf can be computed in R. The following code segment shows R computations of them along with the corresponding χ^2 test. As we have estimated $\hat{p}$, the number of degrees of freedom of the test is $8 - 1 - 1 = 6$.

```
> oc<-c(6,57,206,362,365,256,69,13)
> n<-sum(oc)
> range<-0:7
> phat<-sum(range*oc)/(n*7)
> pmf<-dbinom(range,7,phat)
```

The estimated probability mass function is given in the following code segment.

```
> rbind(range,round(pmf,3))
```

```
       [,1]  [,2]  [,3]  [,4]  [,5]   [,6]   [,7]   [,8]
range 0.000 1.000 2.00  3.000 4.00  5.000  6.000  7.000
      0.006 0.047 0.15  0.265 0.28  0.178  0.063  0.009
```

The p-value is calculated using pchisq with the correct degress of freedom (reduced by one due to the estimation of p).

```
> test.result<-chisq.test(oc,p=pmf)
> pchisq(test.result$statistic,df=6,lower.tail=FALSE)
```

```
X-squared
0.4257546
```

With a p-value $= 0.426$ we would not reject H_0. There is no evidence to refute a binomial probability model for the number of sons in the first seven children of a Swedish minister. The following provides the expected frequencies which can be compared with the the observed.

```
> round(test.result$expected,1)
```

```
[1]    8.5  63.2 200.6 353.7 374.1 237.4  83.7  12.6
```

■

Confidence Intervals

In this section, we have been discussing tests for a discrete random variable with a range consisting of c categories, say, $\{1, 2, \ldots, c\}$. Write the distribution of X as $p_j = p(j) = P(X = j)$, $j = 1, 2, \ldots, c$. Using the notation at the beginning of this section, for a given j, the estimate of p_j is the proportion of sample items in category j; i.e., $\hat{p}_j = O_j/n$. Note that this is a binomial situation where category j is success and all other categories are failures. Hence from expression (2.16), an asymptotic $(1 - \alpha)100\%$ confidence interval for p_j is

$$\hat{p}_j \pm z_{\alpha/2}\sqrt{\frac{\hat{p}_j(1 - \hat{p}_j)}{n}}. \tag{2.24}$$

Another confidence interval of interest in this situation is for a difference in proportions, say, $p_j - p_k$, $j \neq k$. This parameter is the difference in two proportions in a multinomial setting; hence, the standard error[7] of this estimate is

$$\text{SE}(\hat{p}_j - \hat{p}_k) = \sqrt{\frac{\hat{p}_j + \hat{p}_k - (\hat{p}_j - \hat{p}_k)^2}{n}}. \tag{2.25}$$

Thus, an asymptotic $(1 - \alpha)100\%$ confidence interval for $p_j - p_k$ is

$$\hat{p}_j - \hat{p}_k \pm z_{\alpha/2}\sqrt{\frac{\hat{p}_j + \hat{p}_k - (\hat{p}_j - \hat{p}_k)^2}{n}}. \tag{2.26}$$

Example 2.7.2 (Birth Rate of Males to Swedish Ministers, continued). Consider Example 2.7.1 concerning the number of sons in the first seven children of Swedish ministers. Suppose we are interested in the difference in the probabilities of all females or all sons. The following R segment estimates this difference along with a 95% confidence interval, (2.26), for it. The counts for these categories are respectively 6 and 13 with $n = 1334$.

```
> n <- 1334; p0 <- 6/n; p7 <- 13/n
> se <- sqrt((p0+p7-(p0-p7)^2)/n)
> lb <- p0-p7 - 1.96*se; ub <- p0-p7 + 1.96*se
> res<- c(p0-p7,lb,ub)
> res
```

[7]See page 363 of Hogg et al. (2013).

[1] -0.005247376 -0.011645562 0.001150809

Since 0 is in the confidence interval there is no discernible difference in the proportions at level 0.05. ∎

A cautionary note is needed here. In general, many confidence intervals can be formulated for a given situation. For example, if there are c categories then there are $\binom{c}{2}$ possible pairwise comparison confidence intervals. In such cases, the overall confidence may slip. This is called a multiple comparison problem (MCP) in statistics. There are several procedures to use. One such procedure is the **Bonferroni procedure**. Suppose there are m confidence intervals of interest. Then if each confidence interval is obtained with confidence coefficient $(1-(\alpha/m))$, the simultaneous confidence of all of the intervals is at least $1-\alpha$. See Exercise 2.8.25.

2.7.2   Several Discrete Random Variables

A frequent application of goodness-of-fit tests concerns several discrete random variables, say $X_1, \ldots, X_r$, which have the same range $\{1, 2, \ldots, c\}$. The hypotheses of interest are

$$H_0: \qquad X_1, \ldots, X_r \text{ have the same distribution}$$
$$H_A: \quad \text{Distributions of } X_i \text{ and } X_j \text{ differ for some } i \neq j. \qquad (2.27)$$

Note that the null hypothesis does not specify the common distribution. Information consists of independent random samples on each random variable. Suppose the random sample on X_i is of size n_i. let $n = \sum_{i=1}^{r} n_i$ denote the total sample size. The observed frequencies are

$$O_{ij} = \#\{\text{sample items in sample drawn on } X_i \text{ such that } X_i = j\},$$

for $i = 1, \ldots, r$ and $j = 1, \ldots, c$. The set of $\{O_{ij}\}$s form a $r \times c$ matrix of observed frequencies. These matrices are often referred to as **contingency tables**. We want to compare these observed frequencies to the expected frequencies under H_0. To obtain these we need to estimate the common distribution $(p_1, \ldots, p_c)^T$, where p_j is the probability that category j occurs. The nonparametric estimate of p_j is

$$\hat{p}_j = \frac{\sum_{i=1}^{r} O_{ij}}{n}, \quad j = 1, \ldots, c.$$

Hence, the estimated expected frequencies are $\hat{E}_{ij} = n_i \hat{p}_j$. This formula is easy to remember since it is the row total times the column total over the total number. The test statistic is the χ^2-test statistic, (2.23); that is,

$$\chi^2 = \sum_{i=1}^{r} \sum_{j=1}^{c} \frac{(O_{ij} - \hat{E}_{ij})^2}{\hat{E}_{ij}}. \qquad (2.28)$$

TABLE 2.1

Contingency Table for Type of Crime and
Alcoholic Status Data.

Crime	Alcoholic	Non-Alcoholic
Arson	50	43
Rape	88	62
Violence	155	110
Theft	379	300
Coining	18	14
Fraud	63	144

For degrees of freedom, note that each row has $c - 1$ free cells because the
sample sizes n_i are known. Further $c - 1$ estimates had to be made. So there
are $r(c - 1) - (c - 1) = (r - 1)(c - 1)$ degrees of freedom. Thus, an asymptotic
level α test is to reject H_0 if $\chi^2 \geq \chi^2_{\alpha,(r-1)(c-1)}$, where $\chi^2_{\alpha,(r-1)(c-1)}$ is the α
critical value of a χ^2-distribution with $(r - 1)(c - 1)$ degrees of freedom. This
test is often referred to as the χ^2-test of homogeneity (same distributions).
We illustrate it with the following example.

Example 2.7.3 (Type of Crime and Alcoholic Status). The contingency table, Table 2.1, contains the frequencies of criminals who committed certain
crimes and whether or not they are alcoholics. We are interested in seeing
whether or not the distribution of alcoholic status is the same for each type
of crime. The data were obtained from Kendall and Stuart (1979).
To compute the test for homogeneity for this data in R, assume the contingency table is in the matrix `ct`. Then the command is `chisq.test(ct)`, as
the following R session shows:

```
> c1 <- c(50,88,155,379,18,63)
> c2 <- c(43,62,110,300,14,144)
> ct <- cbind(c1,c2)
> chifit <- chisq.test(ct)
> chifit

        Pearson's Chi-squared test

data:  ct
X-squared = 49.7306, df = 5, p-value = 1.573e-09

> (chifit$residuals)^2

            c1          c2
[1,]   0.01617684  0.01809979
[2,]   0.97600214  1.09202023
[3,]   1.62222220  1.81505693
[4,]   1.16680759  1.30550686
```

```
[5,]   0.07191850   0.08046750
[6,]  19.61720859  21.94912045
```

The result is highly significant, but note that most of the contribution to the test statistic comes from the crime fraud. Next, we eliminate fraud and retest.

```
>  ct2 <- ct[-6,]
>  chisq.test(ct2)
```

          Pearson's Chi-squared test

```
data:  ct2
X-squared = 1.1219, df = 4, p-value = 0.8908
```

These results suggest that conditional on the criminal not committing fraud, his alcoholic status and type of crime are independent. ∎

Confidence Intervals

For a given category, say, j, it may be of interest to obtain confidence intervals for differences such as $P(X_i = j) - P(X_{i'} = j)$. In the notation of this section, the estimate of this difference is $(O_{ij}/n_i) - (O_{i'j}/n_{i'})$, where n_i and $n_{i'}$ are the respective sums of rows i and i' of the contingency table. Since the samples on these random variables are independent, the two-sample proportion confidence interval given in expression (2.19) can be used. The cautionary note regarding simultaneous confidence of the last section holds here, also.

2.7.3   Independence of Two Discrete Random Variables

The χ^2 goodness-of-fit test can be used to test the independence of two discrete random variables. Suppose X and Y are discrete random variables with respective ranges $\{1, 2, \ldots, r\}$ and $\{1, 2, \ldots, c\}$. Then we can write the hypothesis of independence between X and Y as

$$
\begin{aligned}
H_0 &: P[X = i, Y = j] &= P[X = i]P[Y = j] \quad \text{for all } i \text{ and } j \text{ versus} \\
H_A &: P[x = i, Y = j] &\neq P[X = i]P[Y = j] \quad \text{for some } i \text{ and } j.
\end{aligned} \quad (2.29)
$$

To test this hypothesis, suppose we have the observed the random sample $(X_1, Y_1), \ldots, (X_n, Y_n)$ on (X, Y). We categorize these data into the $r \times c$ contingency table with frequencies O_{ij} where

$$
O_{ij} = \#_{1 \leq l \leq n}\{(X_l, Y_l) = (i, j)\}.
$$

So the O_{ij}s are our observed frequencies and there are initially $rc - 1$ free cells. The expected frequencies are formulated under H_0. Note that the maximum likelihood estimates (mles) of the marginal distributions of $P[X = i]$ and $P[Y = j]$ are the respective statistics $O_{i\cdot}/n$ and $O_{\cdot j}/n$. Hence, under H_0, the mle of $P[X = i, Y = j]$ is the product of these marginal distributions. So the expected frequencies are

$$\hat{E}_{ij} = n\frac{O_{i.}}{n}\frac{O_{.j}}{n} = \frac{i\text{th row total} \times j\text{th col. total}}{\text{total number}}, \qquad (2.30)$$

which is the same formula as for the expected frequencies in the test for homogeneity. Further, the degrees of freedom are also the same. To see this, there are $rc - 1$ free cells, and to formulate the expected frequencies we had to estimate $r - 1 + c - 1$ parameters. Hence, the degrees of freedom are: $rc - 1 - r - c + 2 = (r - 1)(c - 1)$. Thus the R code for the χ^2-test of independence is the same as for the test of homogeneity. Several examples are given in the exercises.

Confidence Intervals

Notice that the sampling scheme in this section consists of one-sample over $r \times c$ categories. Hence, it is the same scheme as in the beginning of this section, Section 2.7.1. The estimate of each probability $p_{ij} = P[X = i, Y = j]$ is O_{ij}/n and a confidence interval for p_{ij} is given by expression (2.24). Likewise, confidence intervals for differences of the form $p_{ij} - p_{i'j'}$ can be obtained by using expression (2.26).

2.7.4 McNemar's Test

McNemar's test for significant change is used in many applications. The data are generally placed in a contingency table but the analysis is not the χ^2-goodness-of-fit tests discussed earlier. A simple example motivates the test. Suppose A and B are two candidates for a political office who are having a debate. Before and after the debate, the preference, A or B, of each member of the audience is recorded. Given a change in preference of candidate, we are interested in the difference in the change from B to A minus the change from A to B. If the estimate of this difference is significantly greater than 0, we might conclude that A won the debate.

For notation assume we are observing a pair of discrete random variables X and Y. In most applications, the ranges of X and Y have two values, say, $\{0, 1\}$.[8]. In our simple debate example, the common range can be written as $\{\text{For A, For B}\}$. Note that there are four categories $(0, 0), (0, 1), (1, 0), (1, 1)$. Let p_{ij}, $i, j = 0, 1$, denote the respective probabilities of these categories. Consider the hypothesis

$$H_0 : p_{01} - p_{10} = 0 \text{ versus } H_A : p_{01} \neq p_{10}. \qquad (2.31)$$

One-sided tests are of interest, also; for example, in the debate situation, the claim that B wins the debate is expressed by the alternative H_A :

---

[8]See Hettmansperger and McKean (1973) for generalizations to more than two categories.

$p_{01} > p_{10}$. Let $(X_1, Y_1), \ldots, (X_n, Y_n)$ denote a random sample on (X, Y). Let N_{ij}, $i, j = 0, 1$, denote the respective frequencies of the categories $(0, 0), (0, 1), (1, 0), (1, 1)$. For convenience, the data can be written in the contingency table

	0	1
0	N_{00}	N_{01}
1	N_{10}	N_{11}

The estimate of $p_{01} - p_{10}$ is $\hat{p}_{01} - \hat{p}_{10} = (N_{01}/n) - (N_{10}/n)$. This is the difference in two proportions in a multinomial setting; hence, the standard error of this estimate is given in expression (2.26). For convenience, we repeat it with the current notation.

$$\text{SE}(\hat{p}_{01} - \hat{p}_{10}) = \sqrt{\frac{\hat{p}_{01} + \hat{p}_{10} - (\hat{p}_{01} - \hat{p}_{10})^2}{n}}. \tag{2.32}$$

The Wald test statistic is the z-statistic which is the ratio of $\hat{p}_{01} - \hat{p}_{10}$ over its standard error. Usually, though, a scores test is used. In this case the squared difference in the numerator of the standard error is replaced by 0, its parametric value under the null hypothesis. Then the square of the z-scores test statistic reduces to

$$\chi^2 = \frac{(N_{01} - N_{10})^2}{N_{01} + N_{10}}. \tag{2.33}$$

Under H_0, this test statistic has an asymptotic χ^2-distribution with 1 degree of freedom. Letting χ_0^2 be the realized values of the test statistic once the sample is drawn, the p-value of this test is $1 - F_{\chi^2}(\chi_0^2; 1)$. For a one-sided test, simply divide this p-value by 2.

Actually, an exact test is easily formulated. Note that this test is conditioned on the categories $(0, 1)$ and $(1, 0)$. Furthermore, the null hypothesis says that these two categories are equilikely. Hence under the null hypothesis, the statistic N_{01} has a binomial distribution with probability of success $1/2$ and $N_{01} + N_{10}$ trials. So the exact p-value can be determined from this binomial distribution. While either the exact or the asymptotic p-value is easily calculated by R, we recommend the exact p-value.

Example 2.7.4 (Hodgkin's Disease and Tonsillectomy). Hollander and Wolfe (1999) report on a study concerning Hodgkin's disease and tonsillectomy. A theory purports that tonsils offer protection against Hodgkin's disease. The data in the study consist of 85 paired observations of siblings. For each pair, one of the pair have Hodgkin's disease and the other does not. Whether or not each had a tonsillectomy was also reported. The data are:

		Sibling	
		Tonsillectomy (0)	No Tonsillectomy (1)
Hodgkin's	Tonsillectomy (0)	26	15
Patients	No Tonsillectomy (1)	7	37

If the medical theory is correct then $p_{01} > p_{10}$. So we are interested in a one-sided test. The following R calculations show how easily the test statistic and p-value (including the exact) are calculated:

```
> teststat <- (15-7)^2/(15+7)
> pvalue <- (1 - pchisq(teststat,1))/2
> pexact <- 1 - pbinom(14,(15+7),.5)
> c(teststat,pvalue,pexact)

[1] 2.90909091 0.04404076 0.06690025
```

If the level of significance is set at 0.05 then different conclusions may be drawn depending on whether or not the exact p-value is used. ∎

Remark 2.7.1. In practice, the p-values for the χ^2-tests discussed in this section are often the asymptotic p-values based on the χ^2-distribution. For McNemar's test we have the option of an exact p-value based on a binomial distribution. There are other situations where an exact p-value is an option. One such case concerns contingency tables where both margins are fixed. For such cases, **Fisher's exact test** can be used; see, for example, Chapter 2 of Agresti (1996) for discussion. The R function for the analysis is `fisher.test`. One nonparametric example of this test concerns Mood's two-sample median test (e.g. Hettmansperger and McKean 2011: Chapter 2). In this case, Fisher's exact test is based on a hypergeometric distribution. ∎

2.8 Exercises

2.8.1. Verify, via simulation, the level of the `wilcox.test` when sampling from a standard normal distribution. Use $n = 30$ and levels of $\alpha = 0.1, 0.05, 0.01$. Based on the resulting estimate of α, the empirical level, obtain a 95% confidence interval for α.

2.8.2. Redo Exercise 2.8.1 for a t-distribution using 1,2,3,5,10 degrees of freedom.

2.8.3. Redo Example 2.4.1 without a `for` loop and using the `apply` function.

2.8.4. Redo Example 2.3.2 without a `for` loop and using the `apply` function.

2.8.5. Suppose in a poll of 500 registered voters, 269 responded that they would vote for candidate P. Obtain a 90% percentile bootstrap confidence interval for the true proportion of registered voters who plan to vote for P.

2.8.6. For Example 2.3.1 obtain a 90% two-sided confidence interval for the treatment effect.

2.8.7. Write an R function which computes the sign analysis. For example, the following commands compute the statistic S^+, assuming that the sample is in the vector x.

```
xt <- x[x!=0];   nt <- length(xt);   ind <- rep(0,nt);
ind[xt > 0] <-1;   splus <- sum(ind)
```

2.8.8. Calculate the sign test for the nursery school example, Example 2.3.1. Show that the p-value for the one-sided sign test is 0.1445.

2.8.9. The data for the nursery school study were drawn from page 79 of Siegel (1956). In the data table, there is an obvious typographical error. In the 8th set of twins, the score for the the twin that stayed at home is typed as 82 when it should be 62. Rerun the signed-rank Wilcoxon and t-analyses using the typographical error value of 82.

2.8.10. The contaminated normal distribution is frequently used in simulation studies. A standardized variable, X, having this distribution can be written as

$$X = (1 - I_\epsilon)Z + cI_\epsilon Z,$$

where $0 \leq \epsilon < 1$, I_ϵ has a binomial distribution with $n = 1$ and probability of success ϵ, Z has a standard normal distribution, $c > 1$, and I_ϵ and Z are independent random variables. When sampling from the distribution of X, $(1-\epsilon)100\%$ of the time the observations are drawn from a $N(0, 1)$ distribution but $\epsilon 100\%$ of the time the observations are drawn from a $N(0, c^2)$. These later observations are often outliers. The distribution of X is a mixture distribution; see, for example, Section 3.4.1 of Hogg et al. (2013). We say that X has a $CN(c, \epsilon)$ distribution.

> 1. Using the R functions `rbinom` and `rnorm`, write an R function which obtains a random sample of size n from a contaminated normal distribution $CN(c, \epsilon)$.
>
> 2. Obtain samples of size 100 from a $N(0, 1)$ distribution and a $CN(16, 0.25)$ distribution. Form histograms and comparison boxplots of the samples. Discuss the results.

2.8.11. Perform the simulation study of Example 2.3.2 when the population has a $CN(16, 0.25)$ distribution. For the alternatives, select values of θ so the spread in empirical powers of the signed-rank Wilcoxon test ranges from approximately 0.05 to 0.90.

2.8.12. The ratio of the expected squared lengths of confidence intervals is a measure of efficiency between two estimators. Based on a simulation of size 10,000, estimate this ratio between the Hodges–Lehmann and the sample mean for $n = 30$ when the population has a standard normal distribution. Use 95% confidence intervals. Repeat the study when the population has a t-distribution with 2 degrees of freedom.

2.8.13. Suppose the cure rate for the standard treatment of a disease is 0.60. A new drug has been developed for the disease and it is thought that the cure rate for patients using it will exceed 0.60. In a small clinical trial 48 patients having the disease were treated with the new drug and 34 were cured.

(a) Let p be the probability that a patient having the disease is cured by the new drug. Write the hypotheses of interest in terms of p.

(b) Determine the p-value for the clinical study. What is the decision for a nominal level of 0.05?

2.8.14. Let p be the probability of success. Suppose it is of interest to test

$$H_0 : p = 0.30 \text{ versus } H_A : p < 0.30.$$

Let S be the number of successes out of 75 trials. Suppose we reject H_0, if $S \leq 16$.

(a) Determine the significance level of the test.

(b) Determine the power of the test if the true p is 0.25.

(c) Determine the power function for the test for the sequence for the probabilities of success in the set $\{0.02, 0.03, \ldots, 0.35\}$. Then obtain a plot of the power curve.

2.8.15. For the situation of Exercise 2.8.13, a larger clinical study was run. In this study, patients were randomly assigned to either the standard drug or the new drug. Let p_1 and p_2 denote the cure rates for patients under the new drug and the standard drug, respectively. The hypotheses of interest are:

$$H_0 : p_1 = p_2 \text{ versus } H_A : p_1 > p_2.$$

The results of the study are:

Treatment	No. of Patients	No. Cured
New Drug	200	135
Standard Drug	210	130

(a) Determine the p-value of the scores test (2.21). Conclude at the 5% level of significance.

(b) Obtain the 95% confidence interval for $p_1 - p_2$.

2.8.16. Simulate the power of the Wald and scores type two-sample proportions test for the hypotheses

$$H_0 : p_1 = p_2 \text{ versus } H_A : p_1 > p_2.$$

for the following situation. Assume that population 1 is Bernoulli with $p_1 = 0.6$; population 2 is Bernoulli with $p_2 = 0.5$; the level is $\alpha = 0.05$; and $n_1 = n_2 = 50$. Recall that the call `rbinom(m,n,p)` returns m binomial variates with distribution $bin(n, p)$.

2.8.17. In a large city, four candidates (Smith, Jones, Martinelli, and Wagner) are running for Mayor. A poll was conducted by random dialing with the following results:

Smith	Jones	Martinelli	Wagner	Others
442	208	460	180	205

Using a 95% confidence interval, determine if there is a significant difference between the two front runners.

2.8.18. In Example 2.7.1 we tested whether or not a dataset was drawn from a binomial distribution. For this exercise, generate a sample of size $n = 500$ from a truncated Poisson distribution as illustrated with the following R code:

```
x <- rpois(500,3)
x[x >= 8] = 7
```

(a) Obtain a plot of the histogram of the sample.

(b) Obtain an estimate of the sample proportion (`phat<-mean(x/7)`).

(c) Test to see if the sample has a binomial distribution with $n = 7$, (i.e., use the same test as in Example 2.7.1).

2.8.19. Rasmussen (1992) presents the following data on a survey of workers in a large factory on two variables: their feelings concerning a smoking ban (Approve, Do not approve, Not sure) and Smoking status (Never smoked, Ex-smoker, Current smoker). Use the χ^2-test to test the independence of these two variables. Using a post-test analysis, determine what categories contributed heavily to the dependence.

Smoking status	Approval of the smoking ban		
	Approve	Do not approve	Not sure
Never smoked	237	3	10
Ex-smoker	106	4	7
Current smoker	24	32	11

2.8.20. The following data are drawn from Agresti (1996). It concerns the approval ratings of a Canadian prime minister in two surveys. In the first survey, ratings were obtained on 1600 citizens and then in a second survey, six months later, the same citizens were resurveyed. The data are tabled below. Use McNemar's test to see if given a change in attitude toward the prime minister, the probability of going from approval to disapproval is higher than the probability of going from disapproval to approval. Also determine a 95% confidence interval for the difference of these two probabilities.

First survey	Second survey	
	Approve	Disapprove
Approve	794	150
Disapprove	86	570

2.8.21. Even though the χ^2-tests of homogeneity and independence are the same, they are based on different sampling schemes. The scheme for the test of independence is one-sample of bivariate data, while the scheme for the test of homogeneity consists of one-sample from each population. Let C be a contingency table with r rows and c columns. Assume for the test of homogeneity that the rows contain the samples from the r populations. Determine the (large sample) confidence intervals for each of the following parameters under both schemes, where p_{ij} is the probability of cell (i,j) occurring. Write R code to obtain these confidence intervals assuming the input is a contingency table.

1. p_{11}.

2. $p_{11} - p_{12}$.

2.8.22. Mendel's early work on heredity in peas is well known. Briefly, he conducted experiments and the peas could be either round or wrinkled; yellow or green. So there are four possible combinations: RY, RG, WY, WG. If his theory were correct the peas would be observed in a 9:3:3:1 ratio. Suppose the outcome of the experiment yielded the the following observed data

RY	RG	WY	WG
315	108	101	32

Calculate a p-value and comment on the results.

2.8.23. Suppose there are two ways of making widgets: process A and process B. Assume there is a reliable way in which to measure the overall quality of widgets made from either process such that higher value can be measured with some accuracy.

Suppose that a plant has 25 operators and each operator then makes a widget of each type in random order. The results are such that process A has more value than process B for 20 operators, B has more value than A for 3, and the measurements were not different for 2 operators. These data present pretty convincing evidence in favor of Process A. How likely is such a result due to chance if the processes were actually equal in terms of quality?

2.8.24. Conduct a Monte Carlo simulation to approximate the power of the test discussed in Example 2.4.3 when the true $\theta = 1.5$.

2.8.25. Let $0 < \alpha < 1$. Suppose I_1 and I_2 are respective confidence intervals for two parameters θ_1 and θ_2 both with confidence coefficient $1 - (\alpha/2)$; that is,

$$P_{\theta_i}[\theta_i \in I_i] = 1 - \frac{\alpha}{2}, \quad i = 1, 2.$$

Show that the simultaneous confidence for both intervals is at least $1 - \alpha$, i.e.,

$$P_{\theta_1,\theta_2}[\{\theta_1 \in I_1\} \cap \{\theta_2 \in I_2\}] \geq 1 - \alpha.$$

Hint: Use the method of complements and Boole's inequality, $P[A \cup B] \leq$

$P(a) + P(B)$. Extend the argument to m intervals each with confidence coefficient $1 - (\alpha/2)$ to obtain a set of m simultaneous Bonferroni confidence intervals.

3

Two-Sample Problems

In this chapter, we consider two-sample problems. These are simple but often used models in practice. Even in more complicated designs, contrasts of interest often involve two levels. So the ideas discussed here carry over to these designs.

In Sections 3.1–3.2, we discuss the two-sample Wilcoxon procedure. This includes the usual distribution-free rank test as well as the associated rank-based estimation, standard errors of estimates, and confidence intervals for the shift in locations. Wilcoxon procedures are based on the linear score function. Section 3.2.2 discusses rank-based procedures based on normal scores which are optimal if the underlying populations have normally distributed errors. We extend this discussion to general rank scores with a brief consideration of efficiency in Section 3.5. A Hogg-type adaptive scheme for rank score selection is presented in Section 3.6. For all of these generalizations, as shown, the computation of both testing and estimation (including standard errors and confidence intervals) is easily carried out by the R functions in the packages Rfit and npsm.

In Section 3.3, rank-based procedures for the two-sample scale problem are discussed. The Fligner–Kileen rank-based procedure for testing and estimation in this setting is optimal under normality and, unlike the traditional F-test based on variances, it possesses both robustness of validity and power for nonnormal situations. Rank-based procedures for the related Behrens–Fisher problem (Section 3.4) are also considered. As in other chapters, the focus is on the R computation of these rank-based procedures.

3.1 Introductory Example

In this section, we discuss the Wilcoxon rank-based analysis for the two-sample location problem in context of a real example. In this problem, there are two populations which we want to compare. From each population, we have a sample and based on these two samples we want to infer whether or not there is a difference in location between the populations and, if possible, to measure, with standard error, the difference (size of the effect) between the populations. The test component of the analysis is the Wilcoxon two-sample

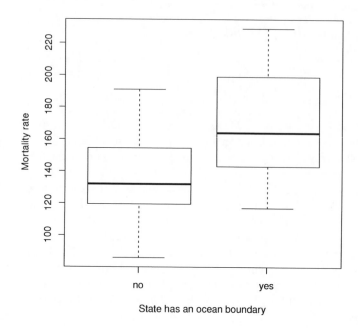

FIGURE 3.1
Mortality rates for white males due to malignant melanoma in the United States.

rank test and the estimation procedure is based on the associated Hodges–Lehmann estimate. We discuss the R computation of the analyses.

Example 3.1.1 (Malignant Melanoma). The dataset USmelanoma, from the package HSAUR2 (Everitt and Hothorn 2014), contains, for each of the 50 states, variables for mortality rate for death due to malignant melanoma of white males, latitude, longitude, and an indicator variable denoting whether the state has an ocean boundary. The primary research questions of interest are: first is there a difference in mortality for ocean states and non-ocean states, and second does the difference prevail after adjusting for latitude and longitude? We address the first question in this chapter and the second question on adjustment in Chapter 5; see Exercise 5.8.13.

In Figure 3.1 we present comparison boxplots for the states that are ocean states (yes) and for the states that are non-ocean states (no). There appears to be an increase in the mortality rate due to malignant melanoma for the ocean states. The difference in the medians is an estimate of the "shift" in mortality rate between the the non-ocean and ocean states.

In Table 3.1, results of the tests and the associated estimates of the size

TABLE 3.1

Estimates of Increase in Mortality Due to Malignant
Melanoma in White Males in the United States.

	Test	p-value	Estimate	Std Error
Least Squares	3.60	0.00	31.49	8.55
Wilcoxon	3.27	0.00	31.00	9.26

effect are displayed for the traditional analysis (two-sample t) and the rank-based analysis (two-sample Wilcoxon). The p-value of the Wilcoxon analysis indicates that there is a significant difference in mortality rates, (from malignant melanoma), between the non-ocean and ocean states. The mortality rate is significantly higher in the ocean states. The Wilcoxon estimate of the shift from non-ocean states to ocean states is about 31 units with standard error 9 units. The results of the traditional LS analysis are similar. ∎

In the next two subsections, we present the Wilcoxon rank-based analysis. We first discuss testing and then estimation.

3.2   Rank-Based Analyses

For the most part, in this book, we are concerned with the rank-based fitting of models and testing hypotheses defined in terms of parameters. The roots of nonparametric methods, though, are distribution free tests, which work well for both the location model, (3.3), as well as for nonparametric settings. In this section, we first briefly discuss these procedures in the general nonparametric setting of stochastic ordering for which these tests are consistent and then discuss them in terms of the location model.

3.2.1   Wilcoxon Test for Stochastic Ordering of Alternatives

Let $X_1, \ldots, X_{n_1}$ denote a random sample from a distribution with the cdf F. Similarly, let $Y_1, \ldots, Y_{n_2}$ denote a random sample from a distribution with the cdf G. Let $n = n_1 + n_2$ denote the total sample size. We begin with the general case which only requires that the response variables be measured on at least an ordinal scale.

The hypotheses of interest are given by

$$H_0 : F(t) = G(t) \text{ vs. } H_A : F(t) \leq G(t), \tag{3.1}$$

where the inequality is strict for at least some t, (in the common support of both X and Y). For the alternative, we often say that X is **stochastically larger** than Y or that X tends to beat Y. This concept is illustrated by the

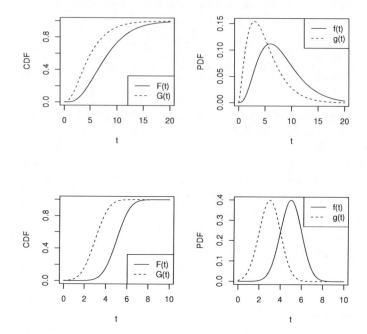

FIGURE 3.2
Plots of stochastic ordering. X is stochastically larger than Y.

cdfs in the left top panel of Figure 3.2. The right panels show the corresponding
pdfs. The bottom panels show a location shift model, which forms a subfamily
of stochastic ordering.

For the discussion in this section, we consider the test based on Wilcoxon
scores. For this test, the samples are combined into one sample and then
ranked from 1 to n, low to high. Let $R(Y_j)$ denote the rank of Y_j in this
combined ranking. Then the Wilcoxon test statistic is

$$T = \sum_{j=1}^{n_2} R(Y_j). \tag{3.2}$$

The command `wilcox.test` is available in base R and we highlight its use in
this section. For the hypotheses (3.1), the Wilcoxon test rejects H_0 for small
values of T. Under H_0 the two samples are from the same population; hence,
any subset of ranks is equilikely as any other subset of the same size. For
example, the probability that a subset of n_2 rankings is assigned to the Y's is
$\binom{n}{n_2}^{-1}$. Thus the null distribution of T is free of the population distribution
and we say that the Wilcoxon test is **distribution-free**. Therefore the p-value

of the Wilcoxon test can be based on the exact distribution of T (by calculating the distribution of the ranks) or by using a large sample approximation.

Example 3.2.1 (Esophageal Cancer). This example is based on the case-control study of esophageal cancer in Ile-et-Vilaine, France (Breslow et al. 1980). These data are available in the `datasets` package. We test the hypothesis that alcohol consumption is the same in the two groups, using as our dataset a sample of cases and controls.

Figure 3.3 displays a stacked bar chart of ordinal data in this study. The following code segment illustrates the creation of the graphic.

```
> library(datasets)
> data(esoph)
> x<-rep(esoph$alcgp,esoph$ncases)
> y<-rep(esoph$alcgp,esoph$ncontrols)
> z<-c(x,y)
> w<-c(rep(1,length(x)),rep(0,length(y)))
> barplot(table(z,w),names.arg=c('Cases','Controls'),
+ legend.text=levels(esoph$alcgp))
```

Below is the output for the Wilcoxon procedure for testing if the case group tends to have higher levels of alcohol consumption than the control group. In this case we use the default settings of a large sample p-value with a continuity correction. Note that we have converted the data from the original ordered categorical (or ordinal) data to numeric, as required by the base R function `wilcox.test`.

```
> x<-as.numeric(x)
> y<-as.numeric(y)
> wilcox.test(x,y)

        Wilcoxon rank sum test with continuity correction

data:  x and y
W = 135611.5, p-value < 2.2e-16
alternative hypothesis: true location shift is not equal to 0
```

The result is highly significant (the cases tend to consume more alcohol). This result is not surprising based on the bar chart. ∎

3.2.2 Analyses for a Shift in Location

We next discuss the Wilcoxon rank-based analysis for the two-sample location problem. Let X and Y be continuous random variables. Let $F(t)$ and $f(t)$ respectively denote the cdf and pdf of X, and $G(t)$ and $g(t)$ respectively denote

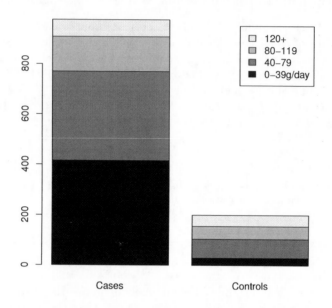

FIGURE 3.3
Stacked bar chart of the esophageal cancer data.

the cdf and pdf of Y. Then we say X and Y follow a **location model**, if for some parameter Δ, $-\infty < \Delta < \infty$,

$$G(t) = F(t - \Delta) \text{ and } g(t) = f(t - \Delta). \tag{3.3}$$

The parameter Δ is the **shift** in location between the random variables Y and X; for instance, it is the difference in medians or in means (provided that the means exist). We can write $\mathcal{L}(Y) = \mathcal{L}(X + \Delta)$, where $\mathcal{L}$ means distribution of (i.e., in law). The location model assumes, in particular, that the scale parameters of X and Y are the same.

Consider two independent samples drawn from the location model. Let $X_1, \ldots, X_{n_1}$ be a random sample on a random variable X with cdf and pdf $F(t)$ and $f(t)$, respectively, and let $Y_1, \ldots, Y_{n_2}$ be a random sample on a random variable Y with cdf and pdf $F(t - \Delta)$ and $f(t - \Delta)$, respectively. Let $n = n_1 + n_2$ denote the total sample size. Assume that the random samples are independent of one another. The hypotheses of interest are

$$H_0 : \Delta = 0 \text{ versus } H_A : \Delta \neq 0. \tag{3.4}$$

One-sided alternatives can also be used. In addition to tests of hypotheses we also discuss estimates of the shift parameter Δ; both point estimates and confidence intervals.

Let $R(Y_i)$ denote the rank of Y_i among the combined samples, i.e., among $X_1, \ldots, X_{n_1}, Y_1, \ldots, Y_{n_2}$. Then the Wilcoxon test statistic is

$$T = \sum_{i=1}^{n_2} R(Y_i). \tag{3.5}$$

There is a second formulation of the Wilcoxon test statistic that is often used. Consider the set of all $n_1 n_2$ differences $\{Y_j - X_i\}$ and let T^+ denote the number of positive differences, i.e.,

$$T^+ = \#_{i,j}\{Y_j - X_i > 0\}. \tag{3.6}$$

We then have the identity

$$T^+ = T - \frac{n_2(n_2 + 1)}{2}. \tag{3.7}$$

The form (3.6) is usually referred to as the Mann–Whitney test statistic.

In the case of tied observations (ties) in the data, one usually assigns the average of the ranks that is allotted to these tied observations. For example, see the following code segment.

```
> z<-c(12,18,11,5,11,5,11,11)
> rank(z)

[1] 7.0 8.0 4.5 1.5 4.5 1.5 4.5 4.5
```

For the two-sided hypothesis (3.4) the test rejects H_0 for small or large values of T. As discussed in the last section, T is distribution-free under H_0 and, hence, exact critical values can be determined from its distribution. If the samples are placed in the R vectors x and y, respectively, the R function `wilcox.test(y,x)` computes T^+ and returns the p-value of the test. By default the p-value is based on the exact distribution for small samples ($n < 50$) and no ties, otherwise it is based on an asymptotic approximation which we discuss below. The argument `exact=FALSE` results in the asymptotic distribution being used and `exact=TRUE` results in the exact distribution being used if no ties are in the data. When obtaining asymptotic p-values, use of a continuity correction may be overridden by setting the argument `correct=FALSE`. The default hypothesis is the two-sided hypothesis. The argument `alternative="less"` (`alternative="greater"`) result in testing the alternative $H_A : \Delta < 0$ ($H_A : \Delta > 0$).

Approximate p-values are based on the asymptotic distribution of T (or T^+). It can be shown that the mean and variance of T under the null hypothesis are given by $n_2(n+1)/2$ and $n_1 n_2(n+1)/12$, respectively. Furthermore, T

has an asymptotic normal distribution. Hence, the standardized test statistic is given by

$$z = \frac{T - [n_2(n+1)/2]}{\sqrt{n_1 n_2(n+1)/12}}. \tag{3.8}$$

In this formulation, p-values are obtained by using the standard normal distribution. In the package npsm we have provided the R function `rank.test` which computes this standardized form. To set these ideas and illustrate the R computation, consider the following simple example.

Example 3.2.2 (Generated t_5-Data). The following are two samples generated from a t-distribution with 5 degrees of freedom. The true shift parameter Δ was set at the value 8. The following code is used to generate the samples.

```
> x<-round(rt(11,5)*10+42,1)
> y<-round(rt(9,5)*10+50,1)
```

The sorted data are

```
> sort(x)

 [1] 29.1 31.1 32.4 34.9 41.0 42.6 45.0 47.9 52.5 59.3 76.6

> sort(y)

 [1] 40.1 45.8 47.2 48.1 49.5 58.3 58.7 62.0 64.8
```

We first obtain the analysis based on the `wilcox.test` function in base R.

```
> wilcox.test(x,y,exact=TRUE)

        Wilcoxon rank sum test

data:  x and y
W = 27, p-value = 0.09518
alternative hypothesis: true location shift is not equal to 0
```

The function `wilcox.test` uses the T^+ (3.6) formulation of the Wilcoxon test statistic, which for these data has the value 27 with p-value 0.0952. Next we obtain the analysis based on the `rank.test` function in npsm.

```
> rank.test(x,y)

statistic =  1.709409 , p-value =  0.08737528
```

The results based on `rank.test` shows that the test statistic is 1.71 with the p-value 0.0874. The results of the exact and asymptotic analyses are quite similar; neither would reject the null hypothesis at a 5% level, while both would be considered significant evidence against the null at a 10% level. Note `wilcox.test(x,y,exact=FALSE,correct=FALSE)` provides the same p-value as `rank.test(x,y)`. ∎

Often an estimate of the shift parameter Δ is desired. The estimator of Δ associated with the Wilcoxon analysis is the Hodges–Lehmann estimator, which is the median of all pairwise differences:

$$\hat{\Delta}_W = \text{med}_{i,j}\{Y_j - X_i\}. \tag{3.9}$$

This estimate is obtained by inverting the Wilcoxon test as we discuss in the next section. Note there are $N_d = n_1 n_2$ pairwise differences between the sample of Ys and the sample of Xs. A distribution-free confidence interval can be constructed from the differences. Let $D_{(1)} < D_{(2)} < \cdots < D_{(N_d)}$ be the ordered differences. If the confidence level is $1 - \alpha$, take c to be the lower $\alpha/2$ critical point of the null distribution of T^+, i.e.,

$$\alpha/2 = P_{H_0}[T^+ \leq c].$$

Then the interval $(D_{(c+1)}, D_{(n-c)})$ is a $(1-\alpha)100\%$ confidence interval for Δ. The asymptotic value for c is given by

$$c = \frac{n_1 n_2}{2} - \frac{1}{2} - z_{\alpha/2}\sqrt{\frac{n_1 n_2(n+1)}{12}},$$

which is rounded to nearest integer. It is generally quite close to the actual value. The R function `wilcox.test` computes the estimate and confidence interval. As an illustration, reconsider Example 3.2.2. The following code segment illustrates the how to obtain a confidence interval for Δ.

```
> wilcox.test(y,x,conf.int=TRUE)

        Wilcoxon rank sum test

data:  y and x
W = 72, p-value = 0.09518
alternative hypothesis: true location shift is not equal to 0
95 percent confidence interval:
 -1.0 18.4
sample estimates:
difference in location
          10.4
```

Note that the confidence interval $(-1, 18.4)$ includes the true Δ which is 8. The point estimate is $\hat{\Delta} = 10.4$. Different confidence levels can be set by changing the value of the input variable `conf.level`. For example, use `conf.level=.99` for a 99% confidence interval.

Note that we can perform the Wilcoxon test equivalently using the following formulation of the Wilcoxon test statistic:

$$T_W = \sum_{i=1}^{n_2} a[R(Y_i)],$$

where $a(i) = \varphi_W[i/(n+1)]$ and $\varphi(u) = \sqrt{12}[u - (1/2)]$. The following identity is easy to obtain:

$$T_W = \frac{\sqrt{12}}{n+1}\left[T - \frac{n_2(n+1)}{2}\right].$$

Hence, the statistic T_W is a linear function of the ranks and, for instance, its z score formulation is exactly the same as that of T. We call the function $\varphi_W(u)$ a score function and, more specifically, the Wilcoxon score function. We also call the related scores, $a_W(i)$, the Wilcoxon scores.

Normal Scores

Certainly functions other than the linear function can be used and are sometimes advantageous to using Wilcoxon scores as they may be more efficient. In Section 3.5 we discuss this, offering an optimality result. Later, in Section 3.6, we present an adaptive scheme which automatically selects an appropriate score. In this section, we discuss the analysis based on the scores which are fully asymptotically efficient if the underlying populations are normally distributed. These scores are call the **normal scores**.

The normal scores are generated by the function.

$$\varphi_{ns}(u) = \Phi^{-1}(u), \tag{3.10}$$

where $\Phi^{-1}(u)$ denotes the inverse of the standard normal cumulative distribution function. Letting $a_{ns}(i) = \varphi_{ns}[i/(n+1)]$, the associated test statistic is

$$T_{ns} = \sum_{i=1}^{n_2} a_{ns}[R(Y_i)].$$

Under H_0 this test statistic is distribution-free. Usually, the associated z-test statistic is used. The mean and variance of T_{ns} under the null hypothesis are 0 and

$$\text{Var}_{H_0}(T_{ns}) = \frac{n_1 n_2}{n-1}\sum_{i=1}^{n} a_{ns}^2(i),$$

respectively. Using these moments, an asymptotic level α test of the hypotheses (3.4) is

$$\text{Reject } H_0 \text{ in favor of } H_A, \text{ if } |z_{ns}| \geq z_{\alpha/2}, \tag{3.11}$$

where

$$z_{ns} = \frac{T_{ns}}{\sqrt{\text{Var}_{H_0}(T_{ns})}}.$$

The z-test formulation is computed by the npsm function `rank.test`. The following code segment displays the results of the normal scores test for the data of Example 3.2.2. Notice the normal scores are called by the argument `scores=nscores`.

```
> rank.test(x,y,scores=nscores)
```

```
statistic =  1.606725 , p-value =  0.1081147
```

The standardized test statistic has the value 1.61 with *p*-value 0.1081.

Estimates and confidence intervals are also available as with `rank.test`. The next section discusses the details of the computation, but we illustrate the use of `rank.test` to obtain these values.

```
> rank.test(x,y,scores=nscores,conf.int=TRUE)
```

```
statistic =  1.606725 , p-value =  0.1081147
  percent confidence interval:
-1.618417 22.41855
Estimate: 10.40007
```

3.2.3 Analyses Based on General Score Functions

Recall in Section 3.2.2 that besides the Wilcoxon scores, we introduced the normal scores. In this section, we define scores in general and discuss them in terms of efficiency. General scores are discussed in Section 2.5 of Hettmansperger and McKean (2011). A set of rank-based scores is generated by a function $\varphi(u)$ defined on the interval $(0,1)$. We assume that $\varphi(u)$ is square-integrable and, without loss of generality, standardize as

$$\int_0^1 \varphi(u)\,du = 0 \text{ and } \int_0^1 \varphi^2(u)\,du = 1. \tag{3.12}$$

The generated scores are then $a_\varphi(i) = \varphi[i/(n+1)]$. Because $\int_0^1 \varphi(u)\,du = 0$, we also may assume that the scores sum to 0, i.e., $\sum_{i=1}^n a[i] = 0$. The Wilcoxon and normal scores functions are given respectively by $\varphi_W(u) = \sqrt{12}[u-(1/2)]$ and $\varphi_{ns}(u) = \Phi^{-1}(u)$ where $\Phi^{-1}(u)$ is the inverse of the standard normal cumulative distribution function.

For general scores, the associated process is

$$S_\varphi(\Delta) = \sum_{j=1}^{n_2} a_\varphi[R(Y_j - \Delta)], \tag{3.13}$$

where $R(Y_j - \Delta)$ denotes the rank of $Y_j - \Delta$ among $X_1,\ldots,X_{n_1}$ and $Y_1 - \Delta,\ldots,Y_{n_2} - \Delta$. A test statistic for the hypotheses

$$H_0: \ \Delta = 0 \text{ versus } H_A: \ \Delta > 0 \tag{3.14}$$

is $S_\varphi = S_\varphi(0)$. Under H_0, the Xs and Ys are identically distributed. It then follows that S_φ is distribution-free. Although exact null distributions of S_φ can be numerically generated, usually the null asymptotic distribution is used. The mean and variance of $S_\varphi(0)$ under the null hypothesis are

$$E_{H_0}[S_\varphi(0)] = 0 \text{ and } \sigma_\varphi^2 = V_{H_0} = \frac{n_1 n_2}{n(n-1)} \sum_{i=1}^n a_\varphi^2(i). \tag{3.15}$$

Furthermore, the null distribution of $S_\varphi(0)$ is asymptotically normal. Hence, our standardized test statistic is

$$z_\varphi = \frac{S_\varphi(0)}{\sigma_\varphi}. \tag{3.16}$$

For the hypotheses (3.14), an asymptotic level α test is to reject H_0 if $z_\varphi \geq z_\alpha$, where z_α is the $(1-\alpha)$ quantile of the standard normal distribution. The two-sided and the other one-sided hypotheses are handled similarly.

The npsm function rank.test computes this asymptotic test statistic z_φ along with the corresponding p-value for all the intrinsic scores found in Rfit. The arguments are the samples, the specified score function, and the value of alternative which is greater, two.sided, or less for respective alternatives $\Delta > 0$, $\Delta \neq 0$, or $\Delta < 0$. For example, the call rank.test(x,y,scores=nscores) compute the normal scores asymptotic test for a two-sided alternative. Other examples are given below.

For a general score function $\varphi(u)$, the corresponding estimator $\hat{\Delta}_\varphi$ of the shift parameter Δ solves the equation

$$S_\varphi(\Delta) \doteq 0. \tag{3.17}$$

It can be shown that the function $S_\varphi(\Delta)$ is a step function of Δ which steps down at each difference $Y_j - X_i$; hence, the estimator is easily computed numerically. The asymptotic distribution of $\hat{\Delta}_\varphi$ is given by

$$\hat{\Delta}_\varphi \text{ has an approximate } N\left(\Delta, \tau_\varphi^2 \sqrt{\frac{1}{n_1} + \frac{1}{n_2}}\right), \tag{3.18}$$

where τ_φ is the scale parameter

$$\tau_\varphi^{-1} = \int_0^1 \varphi(u)\varphi_f(u)\,du \tag{3.19}$$

and

$$\varphi_f(u) = -\frac{f'(F^{-1}(u))}{f(F^{-1}(u))}. \tag{3.20}$$

In practice, τ_φ must be estimated based on the two samples. We recommend the Koul, Sievers, and McKean (1987) estimator which is implemented in Rfit. We denote this estimate by $\hat{\tau}_\varphi$. Thus the standard error of the estimate is

$$\text{SE}(\hat{\Delta}_\varphi) = \hat{\tau}_\varphi \sqrt{\frac{1}{n_1} + \frac{1}{n_2}}. \tag{3.21}$$

Based on (3.18), an approximated $(1-\alpha)100\%$ confidence interval for Δ is

$$\hat{\Delta}_\varphi \pm t_{\alpha/2, n-2}\widehat{\tau_\varphi}\sqrt{\frac{1}{n_1} + \frac{1}{n_2}}, \tag{3.22}$$

where $t_{\alpha/2,n-2}$ is the upper $\alpha/2$ critical value of a t-distribution[1] with $n-2$ degrees of freedom.

3.2.4 Linear Regression Model

Looking ahead to the next chapter, we frame the two-sample location problem as a regression problem. We begin by continuing our discussion of the normal scores rank-based analysis of the previous section. In particular, we next discuss the associated estimate of the shift parameter Δ. The `rfit` function of the R package `Rfit` has the built-in capability to fit regression models with a number of known score functions, including the normal scores. So it is worthwhile to take a few moments to set up the two-sample location model as a regression model. Let $Z = (X_1, \ldots, X_{n_1}, Y_1, \ldots, Y_{n_2})^T$. Let c be a $n \times 1$ vector whose ith component is 0 for $1 \le i \le n_1$ and 1 for $n_1 + 1 \le i \le n = n_1 + n_2$. Then we can write the location model as

$$Z_i = \alpha + c_i \Delta + e_i, \tag{3.23}$$

where $e_1, \ldots, e_n$ are iid with pdf $f(t)$. Hence, we can estimate Δ by fitting this regression model. If we use the method of least squares then the estimate is $\overline{Y} - \overline{X}$. If we use the rank-based fit with Wilcoxon scores then the estimate of Δ is the median of the pairwise differences. If instead normal scores are called for then the estimate is the estimate associated with the normal scores test. The following R segment obtains the rank-based normal scores fit for the data of Example 3.2.2. Notice that `scores=nscores` calls for the normal scores rank-based fit. If this call is deleted, the default Wilcoxon fit is computed instead.

```
> z <- c(x,y); ci <- c(rep(0,length(x)),rep(1,length(y)))
> fitns <- rfit(z ~ ci,scores=nscores)
> coef(summary(fitns))
```

	Estimate	Std. Error	t.value	p.value
(Intercept)	41.800000	4.422388	9.451907	2.111075e-08
ci	9.500033	5.801869	1.637409	1.189079e-01

Hence, the normal scores estimate of Δ is 9.5. Notice that the summary table includes the standard error of the estimate, namely 5.8. Using this, an approximate 95% confidence interval for Δ, based on the upper 0.025 t-critical with 20 degrees of freedom, is $(-2.52, 22.52)$.

We close this section with the Wilcoxon and normal scores analyses of a dataset taken from a preclinical study.

Example 3.2.3 (Quail Data). The data for this problem were extracted from

---

[1] For a discussion on the appropriateness of t-critical values, see McKean and Sheather (1991).

a preclinical study on low density lipids (LDL). Essentially, for this example, the study consisted of assigning 10 quail to a diet containing an active drug compound (treated group), which hopefully reduces LDL, while 20 other quail were assigned to a diet containing a placebo (control group); see Section 2.3 of Hettmansperger and McKean (2011) for a discussion of this preclinical experiment. At the end of a specified time, the LDL levels of all 30 quail were obtained. The data are available in the `quail2` dataset. A comparison boxplot of the data is given in Figure 3.4.

The boxplots clearly show an outlier in the treated group. Further, the plot indicates that the treated group has generally lower LDL levels than the placebo group. Using the `rfit` function, we compute the Wilcoxon and normal scores estimates of Δ and their standard errors for the Wilcoxon and normal scores analyses. For comparison purposes we also present these results for the least squares analysis (LS).

```
> library(Rfit)
> fit<-rfit(ldl~treat,data=quail2)
> coef(summary(fit))
```

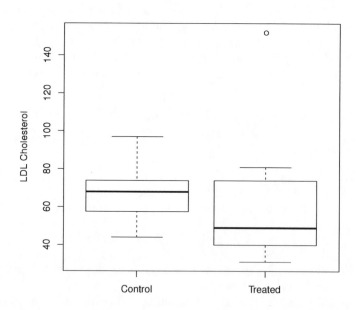

FIGURE 3.4
Comparison boxplots of low density lipids cholesterol.

```
            Estimate Std. Error   t.value       p.value
(Intercept)      65    4.268982 15.226112 4.487132e-15
treat           -14    8.384718 -1.669704 1.061221e-01
```

```
> fitns<-rfit(ldl~treat,data=quail2,scores=nscores)
> coef(summary(fitns))
```

```
            Estimate Std. Error   t.value       p.value
(Intercept) 65.00000   4.116922 15.788494 1.800598e-15
treat      -12.99995   7.669879 -1.694936 1.011828e-01
```

```
> fitls<-lm(ldl~treat,data=quail2)
> coef(summary(fitls))
```

```
            Estimate Std. Error   t value      Pr(>|t|)
(Intercept)     67.2   5.175630 12.9839273 2.270203e-13
treat           -5.0   8.964454 -0.5577585 5.814417e-01
```

The Wilcoxon and normal scores estimates and standard errors are similar. The LS estimate, though, differs substantially from the robust fits. The LS analysis is severely affected by the outlier. Although not significant at the 5% level, (p-value is 0.1061), the Wilcoxon estimate of shift is much closer to the boxplots' difference in medians, -19, than the LS estimate of shift. ∎

3.3 Scale Problem

Besides differences in location, we are often interested in the difference between scales for populations. Let $X_1, \ldots, X_{n_1}$ be a random sample with the common pdf $f[(x - \theta_1)/\sigma_1]$ and $Y_1, \ldots, Y_{n_2}$ be a random sample with the common pdf $f[(y - \theta_2)/\sigma_2]$, where $f(x)$ is a pdf and $\sigma_1, \sigma_2 > 0$. In this section our hypotheses of interest are

$$H_0 : \eta = 1 \text{ versus } H_A : \eta \neq 1, \tag{3.24}$$

where $\eta = \sigma_2/\sigma_1$. Besides discussing rank-based tests for these hypotheses, we also consider the associated estimation of η, along with a confidence interval for η. So here the location parameters θ_1 and θ_2 are nuisance parameters.

As discussed in Section 2.10 of Hettmansperger and McKean (2011), there are asymptotically distribution-free rank-based procedures for this problem. We discuss the Fligner–Killeen procedure based on folded, aligned samples. The aligned samples are defined by

$$\begin{aligned} X_i^* &= X_i - \text{med}\{X_l\}, \quad i = 1, \ldots, n_1 \\ Y_j^* &= Y_j - \text{med}\{Y_l\}, \quad j = 1, \ldots, n_2. \end{aligned} \tag{3.25}$$

Next, the folded samples are $|X_1^*|, \ldots, |X_{n_1}^*|, |Y_1^*|, \ldots, |Y_{n_2}^*|$. The folded samples consist of positive items and their logs, essentially, differ by a location parameter, i.e., $\Delta = \log(\eta)$. This suggests the following log-linear model. Define Z_i by

$$Z_i = \begin{cases} \log|X_i^*| & i = 1, \ldots, n_1 \\ \log|Y_{i-n_1}^*| & i = n_1 + 1, \ldots, n_1 + n_2. \end{cases}$$

Let c be the indicator vector with its first n_1 entries set at 0 and its last n_2 entries set at 1. Then the log-linear model for the aligned, folded sample is

$$Z_i = \Delta c_i + e_i, \quad i = 1, 2, \ldots, n. \tag{3.26}$$

Our rank-based procedure is clear. Select an appropriate score function $\varphi(u)$ and generate the scores $a_\varphi(i) = \varphi[i/(n+1)]$. Obtain the rank-based fit of Model (5.28) and, hence, the estimator $\hat{\Delta}_\varphi$ of Δ. The estimator of η is then $\hat{\eta}_\varphi = \exp\{\hat{\Delta}_\varphi\}$. For specified $0 < \alpha < 1$, denote by (L_φ, U_φ) the confidence interval for Δ based on the fit; i.e.,

$$L_\varphi = \hat{\Delta}_\varphi - t_{\alpha/2}\hat{\tau}_\varphi\sqrt{\frac{1}{n_1} + \frac{1}{n_2}} \text{ and } U_\varphi = \hat{\Delta}_\varphi + t_{\alpha/2}\hat{\tau}_\varphi\sqrt{\frac{1}{n_1} + \frac{1}{n_2}}.$$

An approximate $(1-\alpha)100\%$ confidence interval for η is $(\exp\{L_\varphi\}, \exp\{U_\varphi\})$. Similar to the estimator $\hat{\eta}_\varphi$, an attractive property of this confidence interval is that its endpoints are always positive.

This confidence interval for η can be used to test the hypotheses (3.24); however, the gradient test is often used in practice. Because the log function is strictly increasing, the gradient test statistic is given by

$$S_\varphi = \sum_{j=1}^{n_2} a_\varphi[R(\log(|Y_j^*|))] = \sum_{j=1}^{n_2} a_\varphi[R|Y_j^*|], \tag{3.27}$$

where the ranks are over the combined folded, aligned samples. The standardized test statistic is $z = (S_\varphi - \mu_\varphi)/\sigma_\varphi$, where

$$\mu_\varphi = \frac{1}{n}\sum_{i=1}^n a(i) = \bar{a} \text{ and } \sigma_\varphi^2 = \frac{n_1 n_2}{n(n-1)}\sum_{i=1}^n (a(i) - \bar{a})^2. \tag{3.28}$$

What scores are appropriate? The case of most interest in applications is when the underlying distribution of the random errors in Model (5.28) is normal. In this case the optimal score[2] function is given by

$$\varphi_{FK} = \left(\Phi^{-1}\left(\frac{u+1}{2}\right)\right)^2. \tag{3.29}$$

Hence, the scores are of the form **squared-normal scores**. Note that these are light-tail score functions, which is not surprising because the scores are

---

[2]See Section 2.10 of Hettmansperger and McKean (2011).

optimal for random variables which are distributed as $\log(|W|)$ where W has a normal distribution. Usually the test statistic is written as

$$S_{FK} = \sum_{j=1}^{n_2} \left(\Phi^{-1} \left(\frac{R|Y_j^*|}{2(n+1)} + \frac{1}{2} \right) \right)^2. \tag{3.30}$$

This test statistic is discussed in Fligner and Killeen (1976) and Section 2.10 of Hettmansperger and McKean (2011). The scores generated by the score function (3.29) are in npsm under fkscores. Using these scores, straightforward code leads to the computation of the Fligner–Killeen procedure. We have assembled the code in the R function fk.test which has similar arguments as other standard two-sample procedures.

```
> args(fk.test)
```

```
function (x, y, alternative = c("two.sided", "less", "greater"),
    conf.level = 0.95)
NULL
```

In the call, x and y are vectors containing the original samples (not the folded, aligned samples); the argument alternative sets the hypothesis (default is two-sided); and conf.level sets the confidence coefficient of the confidence interval. The following example illustrates the Fligner–Killeen procedure and its computation.

Example 3.3.1 (Effect of Ozone on Weight of Rats). Draper and Smith (1966) present an experiment on the effect of ozone on the weight gain of rats. Two groups of rats were selected. The control group ($n_1 = 21$) lived in an ozone-free environment for 7 days, while the experimental group ($n_2 = 22$) lived in an ozone environment for 7 days. At the end of the 7 days, the gain in weight of each rat was taken to be the response. The comparison boxplots of the data, Figure 3.5, show a disparity in scale between the two groups.

For this example, the following code segment computes the Fligner–Killeen procedure for a two-sided alternative. The data are in the dataset sievers. We first split the groups into two vectors x and y as input to the function fk.test.

```
> data(sievers)
> x <- with(sievers,weight.gain[group=='Control'])
> y <- with(sievers,weight.gain[group=='Ozone'])
> fk.test(x,y)

statistic =  2.095976 , p-value =  0.03608434
95  percent confidence interval:
1.002458 5.636436
Estimate: 2.377034
```

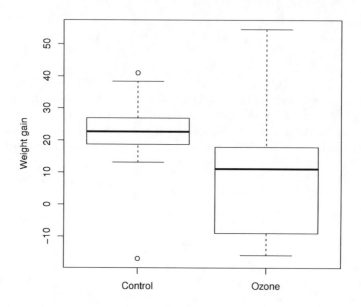

FIGURE 3.5
Comparison boxplots of weight gain in $n_1 = 21$ controls and $n_2 = 22$ ozone treated rats.

Hence, the rank-based estimate of η is 2.337 with the 95% confidence interval of $(1.002, 5.636)$. The standardized Fligner–Killeen test statistic has value 2.09 with the p-value 0.0361 for a two-sided test. Thus there is evidence that rats exposed to ozone have larger variability in their weight gains than nonexposed rats. ∎

The score function (3.29) is optimal for scale if the original samples are from normal populations. Several other score functions are discussed in Hettmansperger and McKean (2011). For example, the Wilcoxon scores are optimal for scale if $|X|$ follows a $F(2, 2)$-distribution.

It is well known that the traditional F-test based on the ratio of sample variances is generally invalid for nonnormal populations. This can be shown theoretically as in Section 2.10.2 of Hettmansperger and McKean (2011). On the other hand, the Fligner–Killeen test is asymptotically distribution-free over all symmetric error pdfs $f(x)$, and, as the next remark discusses, appears to be valid for skewed-contaminated normal distributions. We discuss several pertinent simulation studies next.

Remark 3.3.1 (Simulation Studies Concerning the FK-Test). Conover et al. (1983) discuss the results of a large simulation study of tests for scale in the k-sample problem over many distributions for the random errors. The traditional Bartlett's test (usual F-test in the two-sample problem) is well known to be invalid for nonnormal distributions and this is dramatically shown in this study. Other methods investigated included several folded, aligned tests. One that performed very well uses a test statistic similar to (3.30) except that the exponent is 1 instead of 2. Over symmetric error distributions that test was valid and showed high empirical power but had some trouble with validity for asymmetric distributions. However, in a simulation study conducted by Hettmansperger and McKean (2011), the test based on the test statistic (3.30), (i.e., with the correct exponent 2), (3.30), was empirically valid over a family of contaminated skewed distributions as well as symmetric error distributions.

The base R function `fligner.test` is based on Conover et al. (1983) which used exponent 1, so the test results will differ from `fk.test`. Also, `fk.test` obtains the associated estimate of effect and a confidence interval which are not available in `fligner.test`. ■

In the presence of different scales, one would probably not want to perform the usual analysis for a difference in locations. In the next section, we discuss a rank-based analysis using placements, which is appropriate.

3.4 Placement Test for the Behrens–Fisher Problem

Suppose that we have two populations which differ by location and scale and we are interested in testing that the locations are the same. To be specific, assume for this section that

$$X_1, X_2, \ldots, X_{n_1} \quad \text{is a random sample on } X \text{ with cdf } F(x)$$
$$Y_1, Y_2, \ldots, Y_{n_2} \quad \text{is a random sample on } Y \text{ with cdf } G(x). \quad (3.31)$$

Let θ_X and θ_Y denote the medians of the distributions (populations) $F(x)$ and $G(y)$, respectively. Our two-sided hypothesis of interest is

$$H_0 : \ \theta_X = \theta_Y \text{ versus } H_A \ \theta_X \neq \theta_Y. \quad (3.32)$$

This is called the Behrens–Fisher problem and the traditional test in this situation is the two-sample t-test which uses a t-statistic with the Satterthwaite degrees of freedom correction. This is the default test in the R function `t.test(x,y)`, where x and y are the R vectors containing the samples. In this section, we discuss a version of the two-sample Mann–Whitney–Wilcoxon test which serves as a robust alternative to this approximate t-test.

The nonparametric procedure that we consider was proposed by Fligner and Policello (1981); see, also Section 2.11 of Hettmansperger and McKean (2011) and Section 4.4 of Hollander and Wolfe (1999). It is a modified Mann–Whitney–Wilcoxon test. For its underlying theory, we further assume that the cdfs $F(x)$ and $G(y)$ are symmetric; i.e., symmetric about θ_X and θ_Y, respectively. Provided we have sufficient data, an important diagnostic for checking symmetry is the comparison boxplot of the samples which offers a graphical check of difference in scales and the symmetry of the distributions.

The test statistic is the Mann–Whitney–Wilcoxon statistic defined in expression (3.6), which we rewrite here:

$$T^+ = \#_{i,j}\{Y_j - X_i > 0\} = \sum_{j=1}^{n_2} R(Y_j) - \frac{n_2(n_2 + 1)}{2}.$$

Under the symmetry assumption, $E_{H_0}(T^+) = n_1 n_2/2$. Thus the null expectation of T^+ is the same as in the location problem. The null variance, though, differs from that of the location problem. It is most easily seen in terms of what are known as the **placements**.

Let $P_1, \ldots, P_{n_1}$ denote the placements of the X_is in terms of the Y-sample, which is defined as

$$P_i = \#_j\{Y_j < X_i\}; \quad i = 1, \ldots, n_1. \tag{3.33}$$

In the same way, define the placements of the Y_js in terms of the X-sample as $Q_1, \ldots, Q_{n_2}$ where

$$Q_j = \#_i\{X_i < Y_j\}; \quad j = 1, \ldots, n_2. \tag{3.34}$$

Placements are ranks within a sample, but to avoid confusion with the ranks on the combined samples, the term placement is used. Define

$$\overline{P} = \frac{1}{n_1} \sum_{i=1}^{n_1} P_i \qquad \overline{Q} = \frac{1}{n_2} \sum_{i=1}^{n_2} Q_j$$

$$V_1 = \sum_{i=1}^{n_1} (P_i - \overline{P})^2 \qquad V_2 = \sum_{j=1}^{n_2} (Q_j - \overline{Q})^2. \tag{3.35}$$

Then the standardized test statistic is

$$Z_{fp} = \frac{T^+ - n_1 n_2/2}{(V_1 + V_2 + \overline{PQ})^{1/2}}. \tag{3.36}$$

Under H_0, Z_{fp} has an asymptotic standard normal distribution. So, the Fligner–Policello test of the hypothesis (3.32) is

Reject $H_0 : \theta_X = \theta_Y$ in favor of $H_A : \theta_X \neq \theta_Y$ if $Z_{fp} \geq z_{\alpha/2}$. (3.37)

This test has asymptotically level α.

For computation of this test npsm provides the R function fp.test. Its use is illustrated in the following example.

Example 3.4.1 (Geese). On page 136 of Hollander and Wolfe (1999) a study of healthy and lead-poisoned geese is discussed. The study involved 7 healthy geese and 8 lead-poisoned geese. The response of interest was the amount of plasma glucose in the geese in mg/100 ml of plasma. As discussed in Hollander and Wolfe (1999), the hypotheses of interest are:

$$H_0 : \theta_L = \theta_H \text{ vs. } H_A : \theta_L > \theta_H, \qquad (3.38)$$

where θ_L and θ_H denote the true median plasma glucose values of lead-poisoned geese and healthy geese, respectively. The data are listed below. As shown, the R vectors lg and hg contain respectively the responses of the lead-poisoned and healthy geese. The sample sizes are too small for comparison boxplots, so, instead, the comparison dotplots found in Figure 3.6 are presented. Using Y as the plasma level of the lead-poisoned geese, the test statistic $T^+ = 40$ and its mean under the null hypothesis is 28. So there is some indication that the lead-poisoned geese have higher levels of plasma glucose. The following code segment computes the Fligner–Policello test of the hypotheses, (3.38), based on npsm function fp.test:

```
> lg = c(293,291,289,430,510,353,318)
> hg = c(227,250,277,290,297,325,337,340)
> fp.test(hg,lg,alternative='greater')

statistic =  1.467599 , p-value =  0.07979545
```

Hence, for this study, the Fligner–Policello test finds that lead-poisoned geese have marginally higher plasma glucose levels than healthy geese. ∎

Discussion

In a two-sample problem, when testing for a difference in scale, we recommend the Fligner–Killeen procedure discussed in the last section. Unlike the traditional F-test for a difference in scale, the Fligner–Killeen procedure possesses robustness of validity. Besides the test for scales being the same, the Fligner–Killeen procedure offers an estimate of the ratio of scales and a corresponding confidence interval for this ratio. This gives the user a robust disparity measure (with confidence) for the scales difference in the data, which may be more useful information then an estimate of the difference in locations. For example, based on this information, a user may want to estimate the shift in location after a transformation or to use a weighted estimate of shift based on robust estimates of scale. If, though, a user wants to test for location differences in the presence of difference in scales, then we recommend the robust test offered by the Fligner–Policello procedure.

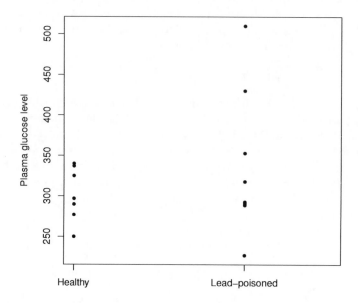

FIGURE 3.6
Comparison dotplots of plasma glucose in healthy and toxic geese.

3.5 Efficiency and Optimal Scores*

In this section, we return to the two-sample location problem, (3.23). We discuss the optimal score function in the advent that the user knows the form of the population distributions. In a first reading, this section can be skipped, but even a brief browsing serves as motivation for the next section on adaptive score selection schemes. Recently Doksum (2013) developed an optimality result for the Hodges–Lehmann estimate based on a rank-based likelihood.

3.5.1 Efficiency

In this subsection, we briefly discuss the robustness and efficiency of the rank-based estimates. This leads to a discussion of optimizing the analysis by the suitable choice of a score function.

When evaluating estimators, an essential property is their robustness. In

Section 2.5, we discussed the influence function of an estimator. Recall that it is a measure of the sensitivity of the estimator to an outlier. We say an estimator is **robust** if its influence function is bounded for all possible outliers.

Consider the rank-based estimator $\widehat{\Delta}_\varphi$ based on the score function $\varphi(u)$. Recall that its standard deviation is $\tau_\varphi/\sqrt{n}$; see expression (3.19). Then for an outlier at z, $\widehat{\Delta}_\varphi$ has the influence function

$$\text{IF}(z; \widehat{\Delta}_\varphi) = -I_X(z) \left(\tau_\varphi \frac{n}{n_1} \right) \varphi[(F(z)] + I_Y(z) \left(\tau_\varphi \frac{n}{n_1} \right) \varphi[(F(z)], \quad (3.39)$$

where $I_X(z)$ is 1 or 0 depending on whether z is thought as being from the sample of Xs or Ys, respectively.[3] The indicator $I_Y(z)$ is defined similarly. For a rank-based estimator, provided its score function $\varphi(u)$ is bounded, its influence function is a bounded function of z; hence, the rank-based estimator is robust. There are a few scores used occasionally in practice, which have unbounded score functions. In particular, the normal scores, discussed in Section 3.5.1, belong to this category; although they are technically robust. In practice, though, generally scores with bounded score functions are used.

In contrast, the influence function for the LS estimator $\overline{Y} - \overline{X}$ of $\widehat{\Delta}$ is given by

$$\text{IF}(z; \widehat{\Delta}_\varphi) = -I_X(z) \left(\frac{n}{n_1} \right) z + I_Y(z) \left(\frac{n}{n_2} \right) z; \quad (3.40)$$

which is an unbounded function of z. Hence, the LS estimator is not robust.

We next briefly consider relative efficiency of estimators of Δ for the two-sample location model (3.23). Suppose we have two estimators, $\widehat{\Delta}_1$ and $\widehat{\Delta}_2$, such that, for $i = 1, 2$, $\sqrt{n}(\widehat{\Delta}_i - \Delta)$ converges in distribution to the $N(0, \sigma_i^2)$ distribution. Then the asymptotic relative efficiency (ARE) between the estimators is the ratio of their asymptotic variances; i.e.,

$$\text{ARE}(\widehat{\Delta}_1, \widehat{\Delta}_2) = \frac{\sigma_2^2}{\sigma_1^2}. \quad (3.41)$$

Note that $\widehat{\Delta}_2$ is more efficient than $\widehat{\Delta}_1$ if this ratio is less than 1. Provided that the estimators are location and scale equivariant, this ratio is invariant to location and scale transformations, so for its computation only the form of the pdf needs to be known. Because they minimize norms, all rank-based estimates and the LS estimate are location and scale equivariant. Also, all these estimators have an asymptotic variance of the form $\kappa^2\{(1/n_1) + (1/n_2)\}$. The scale parameter κ is often called the constant of proportionality. Hence, in the ARE ratio, the design part (the part in braces) of the asymptotic variance cancels out and the ratio simplifies to the ratio of the parameters of proportionality; i.e., the κ^2s.

First, consider the ARE between a rank-based estimator and the LS estimator. Assume that the random errors in Model (3.23) have variance σ^2 and

---

[3]See Chapter 2 of Hettmansperger and McKean (2011).

TABLE 3.2
AREs Among the Wilcoxon (W), Sign (S), and LS Estimators When the
Errors Have a Contaminated Normal Distribution with $\sigma_c = 3$ and
Proportion of Contamination ϵ

	ϵ, (Proportion of Contamination)							
	0.00	0.01	0.02	0.03	0.05	0.10	0.15	0.25
ARE(W, LS)	0.955	1.009	1.060	1.108	1.196	1.373	1.497	1.616
ARE(S, LS)	0.637	0.678	0.719	0.758	0.833	0.998	1.134	1.326
ARE(W, S)	1.500	1.487	1.474	1.461	1.436	1.376	1.319	1.218

pdf $f(t)$. Then, from the above discussion, the ARE between the rank-based
estimator with score generating function $\varphi(u)$ and the LS estimator is

$$\text{ARE}(\widehat{\Delta}_\varphi, \widehat{\Delta}_{LS}) = \frac{\sigma^2}{\tau_\varphi^2}, \tag{3.42}$$

where τ_φ is defined in expression (3.19).

If the Wilcoxon scores are chosen, then $\tau_W = [\sqrt{12} \int f^2(t)\, dt]^{-1}$. Hence,
the ARE between the Wilcoxon and the LS estimators is

$$\text{ARE}(\widehat{\Delta}_W, \widehat{\Delta}_{LS}) = 12\sigma^2 \left[\int f^2(t)\, dt\right]^2. \tag{3.43}$$

If the error distribution is normal then this ratio is 0.955; see, for example,
Hettmansperger and McKean (2011). Thus, the Wilcoxon estimator loses less
than 5% efficiency to the LS estimator if the errors have a normal distribution.

In general, the Wilcoxon estimator has a substantial gain in efficiency over
the LS estimator for error distributions with heavier tails than the normal distribution. To see this, consider a family of contaminated normal distributions
with cdfs

$$F(x) = (1 - \epsilon)\Phi(x) + \epsilon\Phi(x/\sigma_c), \tag{3.44}$$

where $0 < \epsilon < 0.5$ and $\sigma_c > 1$. If we are sampling from this cdf, then $(1 - \epsilon)100\%$ of the time we are sampling from a $N(0, 1)$ distribution while $\epsilon100\%$
of the time we are sampling from a heavier tailed $N(0, \sigma_c^2)$ distribution. For
illustration, consider a contaminated normal with $\sigma_c = 3$. In the first row of
Table 3.2 are the AREs between the Wilcoxon and the LS estimators for a
sequence of increasing contamination. Note that even if $\epsilon = 0.01$, i.e., only 1%
contamination, the Wilcoxon is more efficient than LS.

If sign scores are selected, then $\tau_S = [2f(0)]^{-1}$. Thus, the ARE between
the sign and the LS estimators is $4\sigma^2 f^2(0)$. This is only 0.64 at the normal
distribution, so medians are much less efficient than means if the errors have a
normal distribution. Notice from Table 3.2 that for this mildly contaminated
normal, the proportion of contamination must exceed 10% for the sign estimator to be more efficient than the LS estimator. The third row of the table
contains the AREs between the Wilcoxon and sign estimators. In all of these

situations of the contaminated normal distribution, the Wilcoxon estimator is more efficient than the sign estimator.

If the true pdf of the errors is $f(t)$, then the optimal score function is $\varphi_f(u)$ defined in expression[4] (3.20). These scores are asymptotically efficient; i.e., achieve the Rao–Cramer lower bound asymptotically similar to maximum likelihood estimates. For example, the normal scores are optimal if $f(t)$ is a normal pdf; the sign scores are optimal if the errors have a Laplace (double exponential) distribution; and the Wilcoxon scores are optimal for errors with a logistic distribution. The logistic pdf is

$$f(x) = \frac{1}{b} \frac{\exp\{-[x-a]/b\}}{(1 + \exp\{-[x-a]/b\})^2}, \quad -\infty < x < \infty, \tag{3.45}$$

At the logistic distribution the ARE between the sign scores estimator and the Wilcoxon estimator is $(3/4)$, while at the Laplace distribution this ARE is $(4/3)$. This range of efficiencies suggests a simple family of score functions called the Winsorized Wilcoxons. These scores are generated by a nondecreasing piecewise continuous function defined on $(0,1)$ which is flat at both ends and linear in the middle. As discussed in McKean, Vidmar, and Sievers (1989), these scores are optimal for densities with "logistic" middles and "Laplace" tails. Four typical score functions from this family are displayed in Figure 3.7. If the score function is odd about $1/2$, as in Panel (a), then it is optimal for a symmetric distribution; otherwise, it is optimal for a skewed distribution. Those in Panels (b) and (c) are optimal for distributions which are skewed right and skewed left, respectively. The type in Panel (d) is optimal for light-tailed distributions (lighter than the normal distribution), which occasionally are of use in practice. In `Rfit`, these scores are used via the options: `scores=bentscores4` for the type in Panel (a); `scores=bentscores1` for the type in Panel (b); `scores=bentscores3` for the type in Panel (c); and `scores=bentscores2` for the type in Panel (d).

In the two-sample location model, as well as in all fixed effects linear models, the ARE between two estimators summarizes the difference in analyses based on the estimators. For example, the asymptotic ratio of the squares of the lengths of the confidence intervals based on the two estimators is their ARE. Once we have data, we call the estimated ARE between two estimators the **precision coefficient** or the estimated precision of one analysis over the other. For instance, if we use the two rank-based estimators of Δ based on their respective score functions φ_1 and φ_2 then

$$\text{Precision}(\text{Analysis based on } \hat{\Delta}_1, \text{Analysis based on } \hat{\Delta}_2) = \frac{\hat{\tau}_{\varphi_2}^2}{\hat{\tau}_{\varphi_1}^2}. \tag{3.46}$$

For a summary example of this section, we reconsider the quail data. This time we select an appropriate score function which results in an increase in precision (empirical efficiency) over the Wilcoxon analysis.

---

[4]See Chapter 2 of Hettmansperger and McKean (2011).

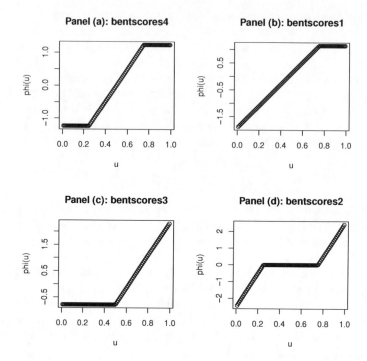

FIGURE 3.7
Plots of four bent score functions, one of each type as described in the text.

Example 3.5.1 (Quail Data, Continued). The data discussed in Example 3.2.3 were drawn from a large study involving many datasets, which was discussed in some detail in McKean et al. (1989). Upon examination of the residuals from many of these models, it appeared that the underlying error distribution is skewed with heavy right-tails, although outliers in the left tail were not usual. Hence, scores of the type `bentscore1` were considered. The final recommended score was of this type with a bend at $u = 3/4$, which is the score graphed in Panel (b) of the above plot. The `Rfit` of the data, using these scores, computes the estimate of shift δ in the following code segment:

```
>     mybentscores = bentscores1
>     mybentscores@param<-c(0.75,-2,1)
>     fit = rfit(z~xmat,scores=mybentscores)
>     summary(fit)
```

Call:
rfit.default(formula = z ~ xmat, scores = mybentscores)

Coefficients:

```
             Estimate Std. Error t.value   p.value
(Intercept)  66.0000    4.1159 16.0355 1.216e-15 ***
xmat        -16.0000    7.6647 -2.0875   0.04606 *
---
Signif. codes:  0 '***' 0.001 '**' 0.01 '*' 0.05 '.' 0.1 ' ' 1

Multiple R-squared (Robust): 0.128132
Reduction in Dispersion Test: 4.11495 p-value: 0.05211
```

The estimate of tau for the bentscores is

```
>     fit$tauhat
```

[1] 19.79027

while the estimate of tau for the Wilcoxon scores is

```
>     rfit(z~xmat)$tauhat
```

[1] 21.64925

In summary, using these bent scores, the rank-based estimate of shift is $\widehat{\Delta} = -16$ with standard error 7.66, which is significant at the 5% level for a two-sided test. Note that the estimate of shift is closer to the estimate of shift based on the boxplots than the estimate based on the Wilcoxon score; see Example 3.2.3. From the code segment, the estimate of the relative precision (3.46) of the bent scores analysis versus the Wilcoxon analysis is $(19.790/21.649)^2 = 0.836$. Hence, the Wilcoxon analysis is about 16% less precise than the bentscore analysis for this dataset.

∎

The use of Winsorized Wilcoxon scores for the quail data was based on an extensive investigation of many datasets. Estimation of the score function based on the rank-based residuals is discussed in Naranjo and McKean (1987). Similar to density estimation, though, large sample sizes are needed for these score function estimators. Adaptive procedures for score selection are discussed in the next section.

As we discussed above, if no assumptions about the distribution can be reasonably made, then we recommend using Wilcoxon scores. The resulting rank-based inference is robust and is highly efficient relative to least squares based inference. Wilcoxon scores are the default in Rfit and are used for most of the examples in this book.

3.6 Adaptive Rank Scores Tests

As discussed in Section 3.5.1, the optimal choice for score function depends on the underlying error distribution. However, in practice, the true distribution is not known and we advise against making any strong distributional assumptions about its exact form. That said, there are times, after working with prior similar data perhaps, that we are able to choose a score function which is appropriate for the data at hand. More often, though, the analyst has a single dataset with no prior knowledge. In such cases, if hypothesis testing is the main inference desired, then the adaptive procedure introduced by Hogg (1974) allows the analyst to first look at the (combined) data to choose an appropriate score function and then allows him to use the same data to test a statistical hypothesis, without inflating the type I error rate.

In this section, we present the version of Hogg adaptive scheme for the two-sample location problem that is discussed in Section 10.6 of Hogg, McKean, and Craig (2013). Consider a situation where the error distributions are either symmetric or right-skewed with tail weights that vary from light-tailed to heavy-tailed. For this situation, we have chosen a scheme that is based on selecting one of the four score functions in Figure 3.8. In applying the adaptive scheme to a particular application, other more appropriate score functions can be selected for the scheme. The number of score functions can also vary.

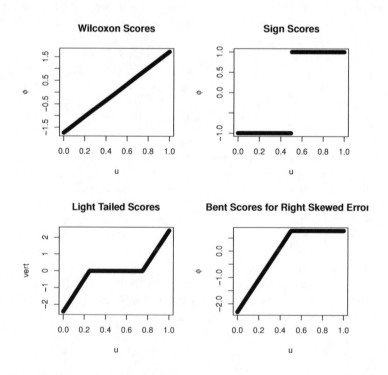

FIGURE 3.8
Score functions used in Hogg's adaptive scheme.

The choice of the score function is made by selector statistics which are based on the order statistics of the combined data. For our scheme, we have chosen two which we label Q_1 and Q_2, which are

$$Q_1 = \frac{\bar{U}_{0.05} - \bar{M}_{0.5}}{\bar{M}_{0.5} - \bar{L}_{0.05}} \quad \text{and} \quad Q_2 = \frac{\bar{U}_{0.05} - \bar{L}_{0.05}}{\bar{U}_{0.5} - \bar{L}_{0.5}}, \quad (3.47)$$

where $U_{0.05}$ is the mean of the Upper 5%, $M_{0.5}$ is the mean of the Middle 50%, and $L_{0.05}$ is the mean of the Lower 5% of the combined sample. Note $Q1$ is a measure of skewness and $Q2$ is a measure of tail heaviness. Benchmarks for the selectors must also be chosen. The following table shows the benchmarks that we used for our scheme.

Benchmark	Score Selected
$Q_2 > 7$	**Sign** Scores
$Q_1 > 1$ & $Q_2 < 7$	**Bent** Score for Right Skewed Data
$Q_1 \leq 1$ & $Q_2 \leq 2$	**Light** Tailed Scores
else	**Wilcoxon**

The R function hogg.test is available in npsm which implements this adaptive method. This function selects the score as discussed above then obtains the test of $H_0 : \Delta = 0$. It returns the test statistic, p-value, and the score selected (boldface names in the above table). We illustrate it on several generated datasets.

Example 3.6.1. The following code generates four datasets and for each the results of the call to function hogg.test is displayed.

```
> m<-50
> n<-55
> # Exponential
> hogg.test(rexp(m),rexp(n))

Hogg's Adaptive Test
Statistic   p.value
  4.32075   0.21434


Scores Selected:  bent

> # Cauchy
> hogg.test(rcauchy(m),rcauchy(n))

Hogg's Adaptive Test
Statistic   p.value
  3.56604   0.31145


Scores Selected:  bent

> # Normal
> hogg.test(rnorm(m),rnorm(n))
```

```
Hogg's Adaptive Test
Statistic    p.value
  2.84318    0.57671
```

```
Scores Selected:  Wilcoxon
```

```
> # Uniform
> hogg.test(runif(m,-1,1),runif(n,-1,1))
```

```
Hogg's Adaptive Test
Statistic    p.value
 -2.1321     0.2982
```

```
Scores Selected:  light
```

∎

Remark 3.6.1. In an investigation of microarray data Kloke et al. (2010) successfully used Hogg's adaptive scheme. The idea is that each gene may have a different underlying error distribution. Using the same score function, for example the Wilcoxon, for each of the genes may result in low power for any number of genes, perhaps ones which would have a much higher power with a more appropriately chosen score function. As it is impossible to examine the distribution of each of the genes in a microarray experiment an automated adaptive approach seems appropriate.

Shomrani (2003) extended Hogg's adaptive scheme to estimation and testing in a linear model. It is based on an initial set of residuals and, hence, its tests are not exact level α. However, for the situations covered in Al-Shomrani's extensive Monte Carlo investigation, the overall liberalness of the adaptive tests was slight. We discuss Al-Shomrani's scheme in Chapter 7. Hogg's adaptive strategy was recently extended to mixed linear models by Okyere (2011). ∎

3.7   Exercises

3.7.1. For the baseball data available in `Rfit`, test the hypothesis of equal heights for pitchers and fielders.

3.7.2. Verify the estimates in Table 3.1.

3.7.3. Verify equivalence of the Mann–Whitney (3.6) and Wilcoxon (3.5) tests for the data in Example 3.2.3.

3.7.4. Let $T = \log S$, where S has an F-distribution with degrees of freedom ν_1 and ν_2. It can be shown, that the distribution of T is left-skewed, right-skewed, or symmetric if $\nu_1 < \nu_2$, $\nu_1 > \nu_2$, or $\nu_1 = \nu_2$, respectively.

(a) Generate two samples of size $n = 20$ from a $\log F(1, 0.25)$ distribution, to one of the two samples add a shift of size $\Delta = 7$. Below is the code to generate the samples.

```
x <- log(rf(20,1,.25))
y <- log(rf(20,1,.25)) + 7.0
```

(b) Obtain comparison dotplots of the samples.

(c) Obtain the LS, Wilcoxon, and bentscores1 estimates of Δ along with their standard errors.

3.7.5. Write R code which generates data as in Exercise 3.7.4. Then run a simulation study to estimate the ARE's among the three estimators: LS, Wilcoxon, and rank-based estimate using the scores bentscores1; where, the empirical ARE between two estimators is their ratio of mean square errors. Use a simulation size of 10,000. Which estimator is best? Which is worst?

3.7.6. Using the function hogg.test, run Hogg's adaptive scheme of Section 3.6 on the data of Exercise 3.7.4.

3.7.7. The truncated normal distribution is implemented in the R package truncnorm (Trautmann et al. 2014). Simulate two independent samples from a truncated normal distribution with parameters $\sigma = 1$, $a = -0.5$, $b = 0.5$; the last two parameters are the min and max respectively. Set the mean for one of the populations to be 5 and the other to be 8.

(a) Obtain comparison boxplots of these two samples.

(b) Using the R function hogg.test provided in the package npsm run the adaptive scheme discussed in Section 3.6 to test for a location difference using these two samples.

3.7.8. Consider the logistic distribution given in expression (3.45). For this exercise, consider the standard pdf with parameters $a = 0$ and $b = 1$.

(a) Show that the cdf is given by

$$F(x) = \frac{1}{1 + e^{-x}}, \quad -\infty < x < \infty.$$

(b) Show that the inverse of the cdf is

$$F^{-1}(u) = \log\left[\frac{u}{1 - u}\right], \quad 0 < u < 1.$$

3.7.9. The logistic distribution is implemented in base R (*logis). Using rlogis obtain a random sample of size 1000 from a logistic distribution with location 0 and scale 1. Use the R function density to obtain a density estimate based on this sample and plot it (e.g. plot(density(rlogis(1000)))). On the graph of the estimated density overlay the graph of the true density (3.45) with parameters $a = 0$ and $b = 1$. Comment on the closeness of the fit.

3.7.10. Write an R function which obtains two independent samples from the logistic distribution, and then runs Hogg's adaptive scheme of Section 3.6 on these data. Have the input to the function be the sample sizes: n and m.

(a) Write R code to simulate Hogg's adaptive scheme for testing for a shift in location when sampling from a logistic. Use sample sizes of $n = 20$ and $m = 25$.

(b) Run your simulation in Part (a) 1000 times. Check the p-value for a 5% test. How many times did the scheme select the Wilcoxon scores? Obtain a 95% confidence interval for the probability of correct selection.

3.7.11. Consider the following two samples. The first was generated from a $N(50, 64)$ distribution while the second was generated from a $N(50, 144)$ distribution. Thus the ratio of scales is 2.0.

Sample from $N(50, 64)$ distribution.								
43.72	58.06	55.57	57.49	64.16	43.49	49.94	49.94	52.58
49.40	38.93	42.65	42.91	46.59	47.88			
Sample from $N(50, 256)$ distribution.								
77.10	42.94	42.32	53.51	66.79	18.26	38.72	29.04	33.54
61.19	32.87	60.98	59.02	60.68	45.11	43.52	35.44	34.31

(a) Obtain comparison boxplots.

(b) Obtain the analysis based on the Fligner–Killeen procedure to test for difference in scales for this data. Was the confidence interval successful in trapping the true ratio of scales given by $\eta = 2.0$?

3.7.12. In Remark 3.3.1, we cited the study by Conover et al. (1983) which shows rather dramatically that the traditional F-test for differences in variances is generally not valid for nonnormal distributions. The double exponential (Laplace) was one of the many distributions over which it was invalid. The pdf of double exponential is given by

$$f(x) = \frac{1}{2}e^{-|x|}, \quad -\infty < x < \infty.$$

(a) Show that the following code generates a sample from this double exponential distribution.

```
rlaplace<-function(n){
    x<-rexp(n)
    ind<-sample(c(-1,1),n,replace=TRUE)
    ind*x
}
```

(b) By running a small simulation study, verify that the Fligner–Killeen procedure is valid for testing for scale differences when sampling from the double exponential distribution. Use sample sizes of $n_1 = 30$ and $n_2 = 35$, and sample under H_0; i.e., draw the samples using the R function rlaplace. Collect the p-values for a two-sided test for 1000 simulations and verify that the test with level 0.05 is valid.

3.7.13. Consider the data of Exercise 3.7.11. The samples were generated from the $N(50, 64)$ and $N(50, 144)$ distributions, respectively. Use the Fligner–Policello test described in Section 3.4 to test for a difference in locations. Recall by Example 3.4.1 that the npsm function fp.test can be used to perform this analysis.

4

Regression I

4.1 Introduction

In this chapter, a nonparametric, rank-based (R) approach to regression modeling is presented. Our primary goal is the estimation of parameters in a linear regression model and associated inferences. As with the previous chapters, we generally discuss Wilcoxon analyses, while general scores are discussed in Section 4.4. These analyses generalize the rank-based approach for the two-sample location problem discussed in the previous chapter. Focus is on the use of the R package Rfit (Kloke and McKean 2012). We illustrate the estimation, diagnostics, and inference including confidence intervals and test for general linear hypotheses. We assume that the reader is familiar with the general concepts of regression analysis.

Rank-based (R) estimates for linear regression models were first considered by Jurečková (1971) and Jaeckel (1972). The geometry of the rank-based approach is similar to that of least squares as shown by McKean and Schrader (1980). In this chapter, the Rfit implementation for simple and multiple linear regression is discussed. The first two sections of this chapter present illustrative examples. We present a short introduction to the more technical aspects of rank-based regression in Section 4.4. The reader interested in a more detailed introduction is referred to Chapter 3 of Hettmansperger and McKean (2011).

We also discuss aligned rank tests in Section 4.5. Using the bootstrap for rank regression is conceptually the same as other types of regression and Section 4.6 demonstrates the Rfit implementation. Nonparametric Smoothers for regression models are considered in Section 4.7. In Section 4.8 correlation is presented, including the two commonly used nonparametric measures of association of Kendall and Spearman. In succeeding chapters we present rank-based ANOVA analyses and extend rank-based fitting to more general models.

4.2   Simple Linear Regression

In this section we present an example of how to utilize `Rfit` to obtain rank-based estimates of the parameters in a simple linear regression problem. Write the simple linear regression model as

$$Y_i = \alpha + x_i\beta + e_i, \quad \text{for } i = 1, \ldots n \tag{4.1}$$

where Y_i is a continuous response variable for the ith subject or experimental unit, x_i is the corresponding value of an explanatory variable, e_i is the error term, α is an intercept parameter, and β is the slope parameter. Interest is on inference for the slope parameter β. The errors are assumed to be iid with pdf $f(t)$. Closed form solutions exist for the least squares (LS) estimates of (4.1). However, in general, this is not true for rank-based (R) estimation.

For the rest of this section we work with an example which highlights the use of `Rfit`. The dataset involved is `engel` which is in the package **quantreg** (Koenker 2013). The data are a sample of 235 Belgian working class households. The response variable is annual household income in Belgian francs and the explanatory variable is the annual food expenditure in Belgian francs.

A scatterplot of the data is presented in Figure 4.1 where the rank-based (R) and least squares (LS) fits are overlaid. Several outliers are present, which affect the LS fit. The data also appear to be heteroscedastic. The following code segment illustrates the creation of the graphic.

```
> library(Rfit)
> data(engel)
> plot(engel)
> abline(rfit(foodexp~income,data=engel))
> abline(lm(foodexp~income,data=engel),lty=2)
> legend("topleft",c('R','LS'),lty=c(1,2))
```

The command `rfit` obtains robust R estimates for the linear regression models, for example (4.1). To examine the coefficients of the fit, use the `summary` command. Critical values and p-values based on a Student t distribution with $n - 2$ degrees of freedom recommended for inference. For this example, `Rfit` used the t-distribution with 233 degrees of freedom to obtain the p-value.

```
> fit<-rfit(foodexp~income,data=engel)
> coef(summary(fit))
```

	Estimate	Std. Error	t.value	p.value
(Intercept)	103.7667620	12.78877598	8.113893	2.812710e-14
income	0.5375705	0.01150719	46.716038	2.621879e-120

Readers with experience modeling in R will recognize that the syntax is similar to using `lm` to obtain a least squares analysis. The fitted regression equation is

$$\widehat{\text{foodexp}} = 103.767 + 0.538 * \text{income}.$$

A 95% confidence interval for the slope parameter (β) is calculated as $0.538 \pm 1.97 * 0.012 = 0.538 \pm 0.023$ or $(0.515, 0.56)$.

Examination of the residuals is an important part of the model building process. The raw residuals are available via the command `residuals(fit)`, though we will focus on Studentized residuals. Recall that Studentized residuals are standardized so that they have an approximate (asymptotic) variance 1. In Figure 4.2, we present a residual plot as well as a normal probability plot of the Studentized residuals. The following code illustrates the creation of the graphic.

```
> rs<-rstudent(fit)
> yhat<-fitted.values(fit)
> par(mfrow=c(1,2))
> qqnorm(rs)
> plot(yhat,rs)
```

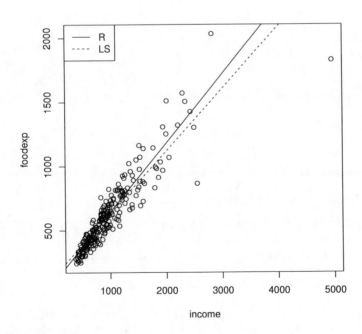

FIGURE 4.1
Scatterplot of Engel data with overlaid regression lines.

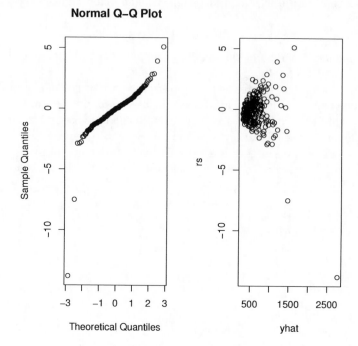

FIGURE 4.2
Diagnostic plots for Engel data.

Outliers and heteroscedasticity are apparent in the residual and normal probability plots in Figure 4.2.

4.3   Multiple Linear Regression

Rank-based regression offers a complete inference for the multiple linear regression. In this section we illustrate how to use `Rfit` to obtain estimates,
standard errors, inference and diagnostics based on an R fit of the model

$$Y_i = \alpha + \beta_1 x_{i1} + \ldots \beta_p x_{ip} + e_i \text{ for } i = 1, \ldots n \qquad (4.2)$$

where $\beta_1, \ldots, \beta_p$ are regression coefficients, Y_i is a continuous response variable, $x_{i1}, \ldots, x_{ip}$ are a set of explanatory variables, e_i is the error term, and
α is an intercept parameter. Interest is on inference for the set of parameters
$\beta_1, \ldots \beta_p$. As in the simple linear model case, for inference, the errors are as-

sumed to be iid with pdf f. Closed form solutions exist for the least squares (LS), however the R estimates must be solved iteratively.

4.3.1 Multiple Regression

In this subsection we discuss a dataset from Morrison (1983: p.64) (c.f. Hettmansperger and McKean 2011). The response variable is the level of free fatty acid (ffa) in a sample of prepubescent boys. The explanatory variables are age (in months), weight (in pounds), and skin fold thickness. In this subsection we illustrate the Wilcoxon analysis and in Exercise 4.9.3 the reader is asked to redo the analysis using bent scores. The model we wish to fit is

$$\text{ffa} = \alpha + \beta_1 \text{age} + \beta_2 \text{weight} + \beta_3 \text{skin} + \text{error}. \qquad (4.3)$$

We use `Rfit` as follows to obtain the R fit of (4.3)

```
> fit<-rfit(ffa~age+weight+skin,data=ffa)
```

and a summary table may be obtained with the `summary` command.

```
> summary(fit)

Call:
rfit.default(formula = ffa ~ age + weight + skin, data = ffa)

Coefficients:
              Estimate Std. Error t.value   p.value
(Intercept)  1.4900402  0.2692512  5.5340 2.686e-06 ***
age         -0.0011242  0.0026348 -0.4267 0.6720922
weight      -0.0153565  0.0038463 -3.9925 0.0002981 ***
skin         0.2749014  0.1342149  2.0482 0.0476841 *
---
Signif. codes:  0 '***' 0.001 '**' 0.01 '*' 0.05 '.' 0.1 ' ' 1

Multiple R-squared (Robust): 0.3757965
Reduction in Dispersion Test: 7.42518 p-value: 0.00052
```

Displayed are estimates, standard errors, and Wald (t-ratio) tests for each of the individual parameters. The variable age is nonsignificant, while weight and skin fold thickness are. In addition there is a robust R^2 value which can be utilized in a manner similar to the usual R^2 value of LS analysis; here $R^2 = 0.38$. Finally a test of all the regression coefficients excluding the intercept parameter is provided in the form of a **reduction in dispersion test**. In this example, we would reject the null hypothesis and conclude that at least one of the nonintercept coefficients is a significant predictor of free fatty acid.

The reduction in dispersion test is analogous to the LS's F-test based on the reduction in sums of squares. This test is based on the reduction (drop) in dispersion as we move from the reduced model (full model constrained by the null hypothesis) to the full model. As an example, for the free fatty acid data, suppose that we want to test the hypothesis:

$$H_0 : \beta_{\text{age}} = \beta_{\text{weight}} = 0 \text{ versus } H_A : \beta_{\text{age}} \neq 0 \text{ or } \beta_{\text{weight}} \neq 0. \quad (4.4)$$

Here, the reduced model contains only the regression coefficient of the predictor skin, while the full model contains all three predictors. The following code segment computes the reduction in dispersion test, returning the F test statistic (see expression (4.15)) and the corresponding p-value:

```
> fitF<-rfit(ffa~age+weight+skin,data=ffa)
> fitR<-rfit(ffa~skin,data=ffa)
> drop.test(fitF,fitR)

Drop in Dispersion Test
F-Statistic      p-value
 1.0768e+01   2.0624e-04
```

The command `drop.test` was designed with the functionality of the command `anova` in traditional analyses in mind. In this case, as the p-value is small, the null hypothesis would be rejected at all reasonable levels of α.

4.3.2 Polynomial Regression

In this section we present an example of a polynomial regression fit using `Rfit`. The data are from Exercise 5 of Chapter 10 of Higgins (2003). The scatterplot of the data in Figure 4.3 reveals a curvature relationship between MPG (`mpg`) and speed (`sp`). Higgins (2003) suggests a quadratic fit and that is how we proceed. That is, we fit the model

$$\text{speed} = \alpha + \beta_1 \text{mpg} + \beta_2 \text{mpg}^2 + \text{error}.$$

To specify a squared term (or any function of the data to be interpreted arithmetically) use the I function:

```
> summary(fit<-rfit(sp~mpg+I(mpg^2),data=speed))

Call:
rfit.default(formula = sp ~ mpg + I(mpg^2), data = speed)

Coefficients:
                Estimate  Std. Error t.value    p.value
(Intercept) 160.7246773   6.4689781 24.8455 < 2.2e-16 ***
mpg          -2.1952729   0.3588436 -6.1176 3.427e-08 ***
I(mpg^2)      0.0191325   0.0047861  3.9975  0.000143 ***
---
```

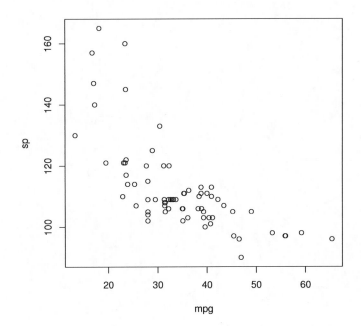

FIGURE 4.3
Scatterplot of miles per gallon vs. top speed.

```
Signif. codes:  0 '***' 0.001 '**' 0.01 '*' 0.05 '.' 0.1 ' ' 1

Multiple R-squared (Robust): 0.5419925
Reduction in Dispersion Test: 46.74312 p-value: 0
```

Note that the quadratic term is highly significant. The residual plot, Figure 4.4, suggests that there is the possibility of heteroscedasticity and/or outliers; however, there is no apparent lack of fit.

Remark 4.3.1 (Model Selection). For rank-based regression, model selection can be performed using forward, backwards, or step-wise procedures in the same way as in ordinary least squares. Procedures for penalized rank-based regression have been developed by Johnson and Peng (2008). In future versions of Rfit we plan to include penalized model selection procedures. ∎

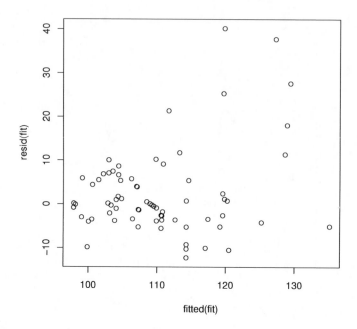

FIGURE 4.4
Residual plot based on the quadratic fit of the Higgins' data.

4.4 Linear Models*

In this section we provide a brief overview of rank-based methods for linear models. Our presentation is by no means comprehensive and is included simply as a convenient reference. We refer the reader interested in a thorough treatment to Chapters 3–5 of Hettmansperger and McKean (2011). This section uses a matrix formulation of the linear model; readers interested in application can skip this and the next section.

4.4.1 Estimation

In this section we discuss rank-based (R) estimation for linear regression models. As is the case with most of the modeling we discuss in this book, the geometry is similar to least squares. Throughout we are interested in estimation and inference on the slope parameters in the following linear model

$$Y_i = \alpha + \beta_1 x_{i1} + \ldots + \beta_p x_{ip} + e_i \text{ for } i = 1, \ldots, n. \tag{4.5}$$

For convenience we rewrite (4.5) as

$$Y_i = \alpha + \boldsymbol{x}_i^T \boldsymbol{\beta} + e_i \text{ for } i = 1, \ldots, n, \tag{4.6}$$

where Y_i is a continuous response variable, $\boldsymbol{x}_i$ is the vector of explanatory variables, α is the intercept parameter, $\boldsymbol{\beta}$ is the vector of regression coefficients, and e_i is the error term. For formal inference, the errors are assumed to be iid with continuous pdf $f(t)$ and finite Fisher information. Additionally, there are design assumptions that are the same as those for the least squares analysis.

Rewrite (4.6) in matrix notation as follows

$$\boldsymbol{Y} = \alpha \boldsymbol{1} + \boldsymbol{X}\boldsymbol{\beta} + \boldsymbol{e} \tag{4.7}$$

where $\boldsymbol{Y} = [Y_1, \ldots, Y_n]^T$ is a $n \times 1$ vector of response variable, $\boldsymbol{X} = [\boldsymbol{x}_1, \ldots, \boldsymbol{x}_n]^T$ is an $n \times p$ design matrix, and $\boldsymbol{e} = [e_1, \ldots, e_n]^T$ is an $n \times 1$ vector of error terms. Recall that the least squares estimator is the minimizer of Euclidean distance between $\boldsymbol{Y}$ and $\hat{\boldsymbol{Y}}_{LS} = \boldsymbol{X}\hat{\boldsymbol{\beta}}_{LS}$. To obtain the R estimator, we use a different measure of distance, Jaeckel's (1972) dispersion function, which is given by:

$$D(\boldsymbol{\beta}) = \|\boldsymbol{Y} - \boldsymbol{X}\boldsymbol{\beta}\|_{\varphi}, \tag{4.8}$$

where $\| \cdot \|_{\varphi}$ is a pseudo-norm defined as

$$\|\boldsymbol{u}\|_{\varphi} = \sum_{i=1}^{n} a(R(u_i))u_i, \tag{4.9}$$

the scores are generated as $a(i) = \varphi\left(\frac{i}{n+1}\right)$, and φ is a nondecreasing **score** function defined on the interval $(0, 1)$. Any of the score functions discussed in the previous chapter for the two-sample location problem can be used in a linear model setting and, therefore, in any of the models used in the remainder of this book. An adaptive procedure for the regression problem is discussed in Section 7.6.

It follows that $D(\boldsymbol{\beta})$, (4.8), is a convex function of $\boldsymbol{\beta}$ and provides a robust measure of distance between $\boldsymbol{Y}$ and $\boldsymbol{X}\boldsymbol{\beta}$. The R estimator of $\boldsymbol{\beta}$ is defined as

$$\hat{\boldsymbol{\beta}}_{\varphi} = \text{Argmin}\|\boldsymbol{Y} - \boldsymbol{X}\boldsymbol{\beta}\|_{\varphi}. \tag{4.10}$$

Note that closed form solutions exist for least squares, however, this is not the case for rank estimation. The R estimates are obtained by minimizing a convex optimization problem. In `Rfit`, the R function `optim` is used to obtain the estimate of $\boldsymbol{\beta}$.

It can be shown, see for example Hettmansperger and McKean (2011), that the solution to (4.10) is consistent with the asymptotically normal distribution given by

$$\hat{\boldsymbol{\beta}}_{\varphi} \dot{\sim} N\left(\boldsymbol{\beta}, \tau_{\varphi}^2 (\boldsymbol{X}^T \boldsymbol{X})^{-1}\right), \tag{4.11}$$

where τ_φ is the scale parameter which is defined in expression (3.19). Note that τ_φ depends on the pdf $f(t)$ and the score function $\varphi(u)$. In Rfit, the Koul, Sievers, and McKean (1987) consistent estimator of τ_φ is computed.

The intercept parameter, α, is estimated separately using a rank-based estimate of location based on the residuals $y_i - \boldsymbol{x}_i^T \hat{\boldsymbol{\beta}}_\varphi$. Generally the median is used, which is the default in Rfit , and which we denote by $\hat{\alpha}$. It follows that $\hat{\alpha}$ and $\hat{\boldsymbol{\beta}}_\varphi$ are jointly asymptotically normal with the variance-covariance matrix

$$\boldsymbol{V}_{\hat{\alpha},\hat{\boldsymbol{\beta}}_\varphi} = \begin{bmatrix} \kappa_n & -\tau_\varphi^2 \overline{\boldsymbol{x}}'(\boldsymbol{X}'\boldsymbol{X})^{-1} \\ -\tau_\varphi^2 (\boldsymbol{X}'\boldsymbol{X})^{-1}\overline{\boldsymbol{x}} & \tau_\varphi^2 (\boldsymbol{X}'\boldsymbol{X})^{-1} \end{bmatrix}, \tag{4.12}$$

where $\kappa_n = n^{-1}\tau_S^2 + \tau_\varphi^2 \overline{\boldsymbol{x}}'(\boldsymbol{X}'\boldsymbol{X})^{-1}\overline{\boldsymbol{x}}$. The vector $\overline{\boldsymbol{x}}$ is the vector of column averages of $\boldsymbol{X}$ and τ_S is the scale parameter[1] $1/[2f(0)]$. The consistent estimator of τ_S, discussed in Section 1.5 of Hettmansperger and McKean (2011), is implemented in Rfit.

4.4.2 Diagnostics

Regression diagnostics are an essential part of the statistical analysis of any data analysis problem. In this section we discuss Studentized residuals. Denote the residuals from the full model fit as

$$\hat{e}_i = Y_i - \hat{\alpha} - \boldsymbol{x}_i \hat{\boldsymbol{\beta}}. \tag{4.13}$$

Then the Studentized residuals are defined as

$$\frac{\hat{e}_i}{s(\hat{e}_i)}$$

where $s(\hat{e}_i)$ is the estimated standard error of $\hat{e}_i$ discussed in Chapter 3 of Hettmansperger and McKean (2011). In Rfit, the command rstudent is used to obtain Studentized residuals from an R fit of a linear model.

4.4.3 Inference

Based on the asymptotic distribution of $\hat{\boldsymbol{\beta}}_\varphi$, (4.11), we present inference for the vector of parameters $\boldsymbol{\beta}$. We discuss Wald type confidence intervals and tests of hypothesis. In addition to these procedures, R-analyses offer the drop in dispersion test which is an analog of the traditional LS test based on the reduction in sums of squares. An estimate of the scale parameter τ_φ is needed for inference and the Koul et al. (1987) estimator is implemented in Rfit.

From (4.11), Wald tests and confidence regions/intervals can easily be obtained. Let $\text{se}(\hat{\beta}_j)$ denote the standard error of $\hat{\beta}_j$. That is $\text{se}(\hat{\beta}_j) = \hat{\tau}_\varphi \left(\boldsymbol{X}^T \boldsymbol{X} \right)_{jj}^{-1/2}$ where $\left(\boldsymbol{X}^T \boldsymbol{X} \right)_{jj}^{-1}$ is the jth diagonal element of $\left(\boldsymbol{X}^T \boldsymbol{X} \right)^{-1}$.

---

[1]See Section 3.5 of Hettmansperger and McKean (2011).

An approximate $(1 - \alpha) * 100\%$ confidence interval for β_j is

$$\hat{\beta}_j \pm t_{1-\alpha/2,n-p-1} \text{se}(\hat{\beta}_j).$$

A Wald test of the hypothesis

$$H_0 : \boldsymbol{M\beta} = \boldsymbol{0} \text{ versus } H_A : \boldsymbol{M\beta} \neq \boldsymbol{0} \tag{4.14}$$

is to reject H_0 if

$$\frac{(\boldsymbol{M}\hat{\boldsymbol{\beta}}_\varphi)^T [\boldsymbol{M}(\boldsymbol{X}^T\boldsymbol{X})^{-1}\boldsymbol{M}^T]^{-1}(\boldsymbol{M}\hat{\boldsymbol{\beta}})/q}{\hat{\tau}_\varphi^2} > F_{1-\alpha,q,n-p-1}$$

where $q = \dim(\boldsymbol{M})$.

Similar to the reduction in the sums of squares test of classical regression, rank-based regression offers a **drop in dispersion** test. Let Ω_F denote the full model space; i.e., the range (column space) of the design matrix $\boldsymbol{X}$ for the full Model (4.7). Let $D(\text{FULL})$ denote the minimized value of the dispersion function when the full model (4.7) is fit. That is, $D(\text{FULL}) = D(\hat{\boldsymbol{\beta}}_\varphi)$. Geometrically, $D(\text{FULL})$ is the distance between the response vector $\boldsymbol{y}$ and the space Ω_F. Let Ω_R denote the reduced model subspace of Ω_F; that is, $\Omega_R = \{\boldsymbol{v} \in \Omega_F : \boldsymbol{v} = \boldsymbol{X\beta} \text{ and } \boldsymbol{M\beta} = \boldsymbol{0}\}$. Let $D(\text{RED})$ denote the minimum value of the dispersion function when the reduced model is fit; i.e., $D(\text{RED})$ is the distance between the response vector $\boldsymbol{y}$ and the space Ω_R. The reduction in dispersion is $RD = D(RED) - D(FULL)$. The drop in dispersion test is a standardization of the reduction in dispersion which is given by

$$F_\varphi = \frac{RD/q}{\hat{\tau}/2}. \tag{4.15}$$

An approximate level α test is to reject H_0 if

$$F_\varphi = \frac{RD/q}{\hat{\tau}/2} > F_{1-\alpha,q,n-p-1}. \tag{4.16}$$

The default ANOVA and ANCOVA `Rfit` analyses described in Chapter 5 use Type III general linear hypotheses (effect being tested is adjusted for all other effects). The Wald type test satisfies this by its formulation. This is true also of the drop in dispersion test defined above; i.e., the reduction in dispersion is between the reduced and full models. Of course, the reduced model design matrix must be computed. This is easily done, however, by using a QR-decomposition of the row space of the hypothesis matrix $\boldsymbol{M}$; see page 210 of Hettmansperger and McKean (2011). For default tests, `Rfit` uses this development to compute a reduced model design matrix for a specified matrix $\boldsymbol{M}$. In general, the `Rfit` function `redmod(xmat,amat)` computes a reduced model design matrix for the full model design matrix `xfull` and the hypothesis matrix `amat`. For traditional LS tests, the corresponding reduction in sums-of-squares is often refereed to as a Type III sums-of-squares.

4.4.4   Confidence Interval for a Mean Response

Consider the general linear model (4.7). Let x_0 be a specified vector of the independent variables. Although we need not assume finite expectation of the random errors, we do in this section, which allows us to use familiar notation. In practice, a problem of interest is to estimate $\eta_0 = E(Y|x_0)$ for a specified vector of predictors x_0. We next consider solutions for this problem based on a rank-based fit of Model (4.7). Denote the rank-based estimates of α and β by $\hat{\alpha}$ and $\hat{\beta}$.

The estimator of η_0 is of course

$$\hat{\eta}_0 = \hat{\alpha} + x_0^T \hat{\beta}. \tag{4.17}$$

It follows that $\hat{\eta}_0$ is asymptotically normal with mean η_0 and variance, using expression (4.12),

$$v_0 = \mathrm{Var}(\hat{\eta}_0) = [1 \ x_0^T] \hat{V}_{\hat{\alpha}, \hat{\beta}_\varphi} \begin{bmatrix} 1 \\ x_0 \end{bmatrix}. \tag{4.18}$$

Hence, an approximate $(1 - \alpha)100\%$ confidence interval for η_0 is

$$\hat{\eta}_0 \pm t_{\alpha/2, n-p-1} \sqrt{\hat{v}_0}, \tag{4.19}$$

where $\hat{v}_0$ is the estimate of v_0, (τ_S and τ_φ are replaced respectively by $\hat{\tau}_S$ and $\hat{\tau}_\varphi$).

Using `Rfit` this confidence interval is easily computed. Suppose `x0` contain the explanatory variables for which we want to estimate the mean response. The following illustrates how to obtain the estimate and standard error, from which a confidence interval can be computed. Assume the full model fit has been obtained and is in `fit` (e.g. `fit<-rfit(y~X)`).

```
x10 <- c(1,x0)
st.err <- sqrt( t(x10)%*%vcov(fit)%*%x10 )
eta0 <- t(x10)%*%fit$coef
```

We illustrate this code with the following example.

Example 4.4.1. The following responses were collected consecutively over time. For convenience take the vector of time to be `t <- 1:15`.

t	1	2	3	4	5	6	7	8
y	0.67	0.75	0.74	-5.57	0.76	2.42	0.16	1.52

t	9	10	11	12	13	14	15
y	2.91	2.74	12.65	5.45	4.81	4.17	3.88

Interest centered on the model for linear trend, $y = \alpha + \beta t + e$, and in estimating the expected value of the response for the next time period $t = 16$. Using Wilcoxon scores and the above code, Exercise 4.9.7 shows that predicted value at time 16 is 5.43 with the 95% confidence interval $(3.38, 7.48)$. Note in practice this might be considered an extrapolation and consideration must be made as to whether the time trend will continue. ∎

There is a related problem consisting of a predictive interval for Y_0 a new (independent of Y) response variable which follows the linear model at x_0. Note that Y_0 has mean η_0. Assume finite variance, σ^2, of the random errors. Then $Y_0 - \hat{\eta}$ has mean 0 and (asymptotic) variance $\sigma^2 + v_0$, where v_0 is given in expression (4.19). If in addition we assume that the random errors have a normal distribution, then we could assume (asymptotically) that the difference $Y_0 - \hat{\eta}$ is normally distributed. Based on these results a predictive interval is easily formed. There are two difficult problems here. One is the assumption of normality and the second is the robust estimation of σ^2. Preliminary Monte Carlo results show that estimation of σ by MAD of the residuals leads to liberal predictive intervals. Also, we certainly would not recommend using the sample variance of robust residuals. Currently, we are investigating a bootstrap procedure.

4.5 Aligned Rank Tests*

An aligned rank test is a nonparametric method which allows for adjustment of covariates in tests of hypotheses. In the context of a randomized experiment to assess the effect of some intervention one might want to adjust for baseline covariates in the test for the intervention. In perhaps the simplest context of a two-sample problem, the test is based on the Wilcoxon rank sum from the residuals of a robust fit of a model on the covariates. Aligned rank tests were first developed by Hodges and Lehmann (1962) for use in randomized block designs. They were developed for the linear model by Adichie (1978); see also Puri and Sen (1985) and Chiang and Puri (1984). Kloke and Cook (2014) discuss aligned rank tests and consider an adaptive scheme in the context of a clinical trial.

For simplicity, suppose that we are testing a treatment effect and each subject is randomized to one of k treatments. For this section consider the model

$$Y_i = \alpha + \boldsymbol{w}_i^T \boldsymbol{\Delta} + \boldsymbol{x}_i^T \boldsymbol{\beta} + e_i \qquad (4.20)$$

where $\boldsymbol{w}_i$ is a $(k-1) \times 1$ incidence vector denoting the treatment assignment for the ith subject, $\boldsymbol{\Delta} = [\Delta_2, \ldots, \Delta_K]^T$ is a vector of unknown treatment effects, $\boldsymbol{x}_i$ is a $p \times 1$ vector of (baseline) covariates, $\boldsymbol{\beta}$ is a vector of unknown regression coefficients, and e_i denotes the error term. The goal of the experiment is to test

$$H_0 : \boldsymbol{\Delta} = \boldsymbol{0}. \qquad (4.21)$$

In this section we focus on developing an aligned rank tests for Model (4.20). We write the model as

$$\boldsymbol{Y} = \alpha \boldsymbol{1} + \boldsymbol{W}\boldsymbol{\Delta} + \boldsymbol{X}\boldsymbol{\beta} + \boldsymbol{e} = \alpha \boldsymbol{1} + \boldsymbol{Z}\boldsymbol{b} + \boldsymbol{e}. \qquad (4.22)$$

Then the full model gradient is

$$S(b) = Z^T a(R(Y - Zb)).$$

First, fit the reduced model

$$Y = \alpha 1 + X\beta + e.$$

Then plug the reduced model estimate $\hat{b}_r = [0^T \hat{\beta}_r^T]^T$ into the full model:

$$S(\hat{b}_r) \doteq \left[\begin{array}{c} W^T a(R(Y - X\hat{\beta}_r)) \\ 0 \end{array} \right].$$

Define $\hat{S}_1 = W^T a(R(Y - X\hat{\beta}_r))$ as the first $k - 1$ elements of $S(\hat{b}_r)$. Then the aligned rank test for (4.21) is based on the test statistic

$$\hat{S}_1^T [W^T W - W^T H_X W]^{-1} \hat{S}_1 = \hat{S}_1^T [W^T H_{X^\perp} W]^{-1} \hat{S}_1$$

where H_X is the projection matrix onto the space spanned by the columns of X. For inference, this test statistic should compared to χ^2_{K-1} critical values.

In the package npsm, we have included the function aligned.test which performs the aligned rank test.

A simple simulated example illustrates the use of the code.

```
> k<-3    # number of treatments
> p<-2    # number of covariates
> n<-10   # number of subjects per treatment
> N<-n*k # total sample size
> y<-rnorm(N)
> x<-matrix(rnorm(N*p),ncol=p)
> g<-rep(1:k,each=n)
> aligned.test(x,y,g)

statistic =  1.083695 , p-value =   0.5816726
```

4.6   Bootstrap

In this section we illustrate the use of the bootstrap for rank-based (R) regression. Bootstrap approaches for M estimation are discussed in Fox and Weisberg (2011) and we take a similar approach. While it is not difficult to write bootstrap functions in R, we make use of the boot library. Our goal is not a comprehensive treatment of the bootstrap, but rather we present an example to illustrate its use.

In this section we utilize the baseball data, which is a sample of 59 professional baseball players. For this particular example we regress the weight of a baseball player on his height.

To use the `boot` function, we first define a function from which `boot` can calculate bootstrap estimates of the regression coefficients:

```
> boot.rfit<-function(data,indices) {
+      data<-data[indices,]
+      fit<-rfit(weight~height,data=data,tau='N')
+      coefficients(fit)[2]
+ }
```

Next use `boot` to obtain the bootstrap estimates, etc.

```
> bb.boot<-boot(data=baseball,statistic=boot.rfit,R=1000)
> bb.boot
```

```
ORDINARY NONPARAMETRIC BOOTSTRAP


Call:
boot(data = baseball, statistic = boot.rfit, R = 1000)


Bootstrap Statistics :
     original      bias     std. error
t1* 5.714278 -0.1575826    0.7761589
```

Our analysis is based on 1000 bootstrap replicates.

Figure 4.5 shows a histogram of the bootstrap estimates and a normal probability plot.

```
> plot(bb.boot)
```

Bootstrap confidence intervals are obtained using the `boot.ci` command. In this segment we obtain the bootstrap confidence interval for the slope parameter.

```
> boot.ci(bb.boot,type='perc',index=1)
```

```
BOOTSTRAP CONFIDENCE INTERVAL CALCULATIONS
Based on 1000 bootstrap replicates

CALL :
boot.ci(boot.out = bb.boot, type = "perc", index = 1)


Intervals :
Level     Percentile
95%    ( 3.75,  7.00 )
Calculations and Intervals on Original Scale
```

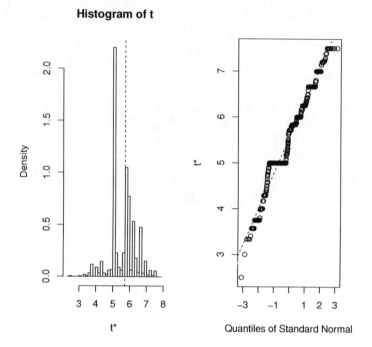

FIGURE 4.5
Bootstrap plots for R regression analysis for modeling weight versus height of
the baseball data.

Exercise 4.9.4 asks the reader to compare the results with those obtained using
large sample inference.

4.7 Nonparametric Regression

In this chapter we have been discussing linear models. Letting Y_i and $x_i = [x_{i1}, \ldots, x_{ip}]^T$ denote the ith response and its associated vector of explanatory variables, respectively, these models are written as

$$Y_i = \alpha + \beta_1 x_{i1} + \beta_2 x_{i2} + \cdots + \beta_p x_{ip} + e_i, \quad i = 1, \ldots, n. \tag{4.23}$$

Note that these models are linear in the regression parameters $\beta_j \; j = 1, \ldots p$; hence, the name linear models. Next consider the model

$$Y_i = \beta_1 \exp\{\alpha + \beta_2 x_{i1} + \beta_3 x_{i3}\} + e_i, \quad i = 1, \ldots, n. \tag{4.24}$$

This model is not linear in the parameters and is an example of a nonlinear model. We explore such models in Section 7.7. The form of Model (4.24) is still known. What if, however, we do not know the functional form? For example, consider the model

$$Y_i = g(x_i) + e_i, \quad i = 1, \ldots, n, \tag{4.25}$$

where the function g is unknown. This model is often called a **nonparametric regression model**. There can be more than one explanatory variable x_i, but in this text we only consider one predictor. As with linear models, the goal is to fit the model. The fit usually shows local trends in the data, finding peaks and valleys which may have practical consequences. Further, based on the fit, residuals are formed to investigate the quality of fit. This is the main topic of this section. Before turning our attention to nonparametric regression models, we briefly consider polynomial models. We consider the case of unknown degree, so, although they are parametric models, they are not completely specified.

4.7.1 Polynomial Models

Suppose we are willing to assume that $g(x)$ is a sufficiently smooth function. Then by Taylor's Theorem, a polynomial may result in a good fit. Hence, consider polynomial models of the form

$$Y_i = \alpha + \beta_1(x_i - \overline{x}) + \beta_2(x_i - \overline{x})^2 + \cdots + \beta_p(x_i - \overline{x})^p + e_i, \quad i = 1, \ldots, n, \tag{4.26}$$

Here x is centered as shown in the model. A disadvantage of this model is that generally the degree of the polynomial is not known. One way of dealing with this unknown degree is to use residual plots based upon iteratively fitting polynomials of different degrees to determine a best fit; see Exercise 4.9.10 for such an example.

To determine the degree of a polynomial model, Graybill (1976) suggested an algorithm based on testing for the degree. Select a large (super) degree P which provides a satisfactory fit of the model. Then set $p = P$, fit the model, and test $\beta_p = 0$. If the hypothesis is rejected, stop and declare p to be the degree. If not, replace p with $p - 1$ and reiterate the test. Terpstra and McKean (2005) discuss the results of a small simulation study which confirmed the robustness of the Wilcoxon version of this algorithm. The npsm package contains the R function polydeg which performs this Wilcoxon version. We illustrate its use in the following example based on simulated data. In this section, we often use simulated data to check how well the procedures fit the model.

Example 4.7.1 (Simulated Polynomial Model). In this example we simulated data from the the polynomial $g(x) = 10 - 3x - 3x^2 + x^3$. One-hundred x values were generated from a $N(0, 3)$-distribution while the added random noise, e_i, was generated from a t-distribution with 2 degrees of freedom and with a multiplicative scale factor of 15. The data are in the set poly. The next code segment shows the call to `polydeg` and the resulting output summary of each of its steps, consisting of the degree tested, the drop in dispersion test statistic, and the p-value of the test. Note that we set the super degree of the polynomial to 5.

```
> deg<-polydeg(x,y,5,.05)
> deg$coll

 Deg    Robust F    p-value
  5    2.331126 0.1301680
  4    0.474083 0.4927928
  3 229.710860 0.0000000

> summary(deg$fitf)

Call:
rfit.default(formula = y ~ xmat)

Coefficients:
             Estimate Std. Error t.value    p.value
(Intercept)  9.501270   3.452188  2.7522   0.007077 **
xmatxc      -7.119670   1.247642 -5.7065 1.279e-07 ***
xmat        -1.580433   0.212938 -7.4220 4.651e-11 ***
xmat         1.078679   0.045958 23.4707 < 2.2e-16 ***
---
Signif. codes:  0 '***' 0.001 '**' 0.01 '*' 0.05 '.' 0.1 ' ' 1

Multiple R-squared (Robust): 0.7751821
Reduction in Dispersion Test: 110.3375 p-value: 0
```

Note that the routine determined the correct degree; i.e., a cubic. Based on the summary of the cubic fit, the 95% confidence interval for the leading coefficient β_3 traps its true value of 1. To check the linear and quadratic coefficients, the centered polynomial must be expanded. Figure 4.6 displays the scatterplot of the data overlaid by the rank-based fit of the cubic polynomial and the associated Studentized residual plot. The fit appears to be good which is confirmed by the random scatter of the Studentized residual plot. This plot also identifies several large outliers in the data, as expected, because the random errors follow a t-distribution with 2 degrees of freedom. ∎

4.7.2   Nonparametric Regression

There are situations where polynomial fits will not suffice; for example, a dataset with many peaks and valleys. In this section, we turn our attention to the nonparametric regression model (4.25) and consider several nonparametric procedures which fit this model. There are many references for nonparametric

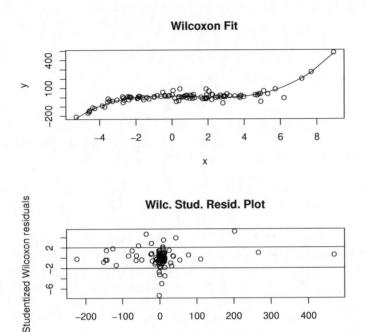

FIGURE 4.6
For the polynomial data of Example 4.7.1: The top panel displays the scatter-plot overlaid with the Wilcoxon fit and the lower panel shows the Studentized residual plot.

regression models. An informative introduction, using R, is Chapter 11 of Faraway (2006); for a more authoritative account see, for example, Wood (2006). Wahba (1990) offers a technical introduction to smoothing splines using reproducing kernel Hilbert spaces.

Nonparametric regression procedures fit local trends producing, hopefully, a smooth fit. Sometimes they are called **smoothers**. A simple example is provided by a running average of size 3. In this case the fit at x_i is $(Y_{i-1} + Y_i + Y_{i+1})/3$. Due to the non-robustness of this fit, often the mean is replaced by the median.

The moving average can be thought of as a weighted average using the discrete distribution with mass sizes of $1/3$ for the weights. This has been generalized to using continuous pdfs for the weighting. The density function used is called a **kernel** and the resulting fit is called a **kernel** nonparametric regression estimator. One such kernel estimator, available in base R, is the

Nadaraya–Watson estimator which is defined at x by

$$\hat{f}_h(x) = \frac{1}{nh} \sum_{i=1}^{n} \frac{w_i}{\sum_{j=1}^{n} w_j} Y_i, \qquad (4.27)$$

where the weights are given by

$$w_i = \frac{1}{h} K\left(\frac{x - x_i}{h}\right) \qquad (4.28)$$

and the kernel $K(x)$ is a continuous pdf. The parameter h is called the bandwidth. Notice that h controls the amount of smoothing. Values of h too large often lead to overly smoothed fits, while values too small lead to overfitting (a jagged fit). Thus, the estimator (4.27) is quite sensitive to the bandwidth. On the other hand, it is generally not as sensitive to the choice of the kernel function. Often, the normal kernel is used. An R function that obtains the Nadaraya–Watson smoother is the function `ksmooth`.

The following code segment obtains the `ksmooth` fit for the polynomial dataset of Example 4.7.1. In the first fit (top panel of Figure 4.7) we used the default bandwidth of $h = 0.5$, while in the second fit (lower panel of the figure) we set h at 0.10. We omitted the scatter of data to show clearly the sensitivity of the estimator (smooth) to the bandwidth setting. For both fits, in the call, we requested the normal kernel.

```
> par(mfrow=c(2,1))
> plot(y~x,xlab=expression(x),ylab=expression(y),pch=" ")
> title("Bandwidth 0.5")
> lines(ksmooth(x,y,"normal",0.5))
> plot(y~x,xlab=expression(x),ylab=expression(y),pch=" ")
> lines(ksmooth(x,y,"normal",0.10))
> title("Bandwidth 0.10")
```

The fit with the smaller bandwidth, 0.10, is much more jagged. The fit with the default bandwidth shows the trend, but notice the "artificial" valley it detected at about $x = 6.3$. In comparing this fit with the Wilcoxon cubic polynomial fit this valley is due to the largest outlier in the data, (see the Wilcoxon residual plot in Figure 4.6). The Wilcoxon fit was not impaired by this outlier.

The sensitivity of the fit to the bandwidth setting has generated a substantial amount of research on data-driven bandwidths. The R package `sm` (Bowman and Azzalini 2014) of nonparametric regression and density fits developed by Bowman and Azzalini contain such data-driven routines; see also Bowman and Azzalini (1997) for details. These are also kernel-type smoothers. We illustrate its computation with the following example.

Example 4.7.2 (Sine Cosine Model). For this example we generated $n = 197$ observations from the model

$$y_i = 5\sin(3x) + 6\cos(x/4) + e_i, \qquad (4.29)$$

Bandwidth 0.5

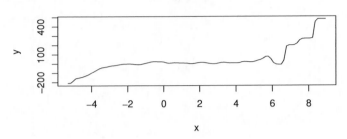

Bandwidth 0.10

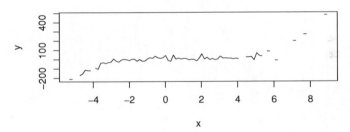

FIGURE 4.7

For the polynomial data of Example 4.7.1: The top panel displays the scatter-plot overlaid with the `ksmooth` fit using the bandwidth set at 0.5 (the default value) while the bottom panel shows the fit using the bandwidth set at 0.10.

where e_i are $N(0, 100)$ variates and x_i goes from 1 to 50 in increments of 0.25. The data are in the set `sincos`. The appropriate `sm` function is `sm.regression`. It has an argument `h` for bandwidth, but if this is omitted a data-driven bandwidth is used. The fit is obtained as shown in the following code segment (the vectors `x` and `y` contain the data). The default kernel is the normal pdf and the option `display="none"` turns off the automatic plot. Figure 4.8 displays the data and the fit.

```
> library(sm)
> fit <- sm.regression(x,y,display="none")
> fit$h              ##    Data driven bandwidth

[1] 4.211251

> plot(y~x,xlab=expression(x),ylab=expression(y))
> with(fit,lines(estimate~eval.points))
> title("sm.regression Fit of Sine-Cosine Data")
```

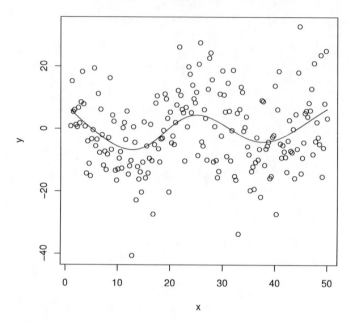

FIGURE 4.8
Scatterplot overlaid with `sm.regression` for the data of Example 4.7.2.

Note that the procedure estimated the bandwidth to be 4.211. ∎

The smoother `sm.regression` is not robust. As an illustration, consider the sine-cosine data of Example 4.7.2. We modified the data by replacing y_{187} with the outlying value of 800. As shown in Figure 4.9, the `sm.regression` fit (solid line) of the modified data is severely impaired in the neighborhood of the outlier. The valley at $x = 36$ has essentially been missed by the fit. For ease of comparison, we have also displayed the `sm.regression` fit on the original data (broken line), which finds this valley.

A smoother which has robust capabilities is `loess` which was developed by Cleveland et al. (1992). This is a base R routine. Briefly, `loess` smooths at a point x_i via a local linear fit. The percentage of data used for the local fit is the analogue of the bandwidth parameter. The default percentage is 75%, but it can be changed by using the argument `span`. Also, by default, the local fitting procedure is based on a weighted least squares procedure. As shown by the broken line fit in Figure 4.10, for the modified sine-cosine data, `loess` is affected in the same way as the sm fit; i.e., it has missed the valley at $x = 36$. Setting the argument `family="symmetric"` in the call to

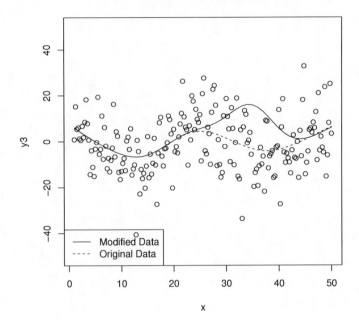

FIGURE 4.9
Scatterplot of modified data ($y_{137} = 800$) overlaid with `sm.regression` fit
(solid line). For comparison, the `sm.regression` fit (broken line) on the original data is also shown.

`loess`, though, invokes a robust local linear model fit. This is the solid line
in Figure 4.10 which is very close to the `sm` fit based on the original data.
Exercise 4.9.18 asks the reader to create a similar graphic. The following code
segment generates Figure 4.10; the figure demonstrates the importance in
obtaining a robust fit in addition to a traditional fit. We close this section
with two examples using real datasets.

Example 4.7.3 (Old Faithful). This dataset concerns the eruptions of Old
Faithful, which is a geyser in Yellowstone National Park, Wyoming. The
dataset is `faithful` in base R. As independent and dependent variables, we
chose respectively the duration of the eruption and the waiting time between
eruptions. In the top panel of Figure 4.11 we display the data overlaid with
the `loess` fit. There is an increasing trend, i.e., longer eruption times lead to
longer waiting times between eruptions. There appear to be two groups in the
plot based on lower or higher duration of eruption times. As the residual plot
shows, the loess fit has detrended the data. ∎

loess Fits of Sine–Cosine Data

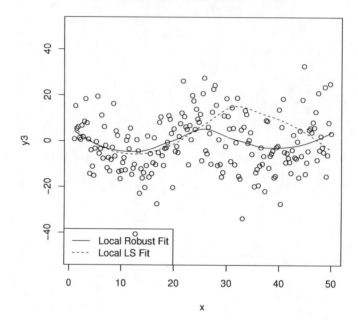

FIGURE 4.10
Scatterplot of modified data ($y_{137} = 800$) overlaid with `loess` fit (solid line).
For comparison, the `sm.regression` fit (broken line) on the original data is
also shown. Note that for clarity, the outlier is not shown.

Example 4.7.4 (Maximum January Temperatures in Kalamazoo). The
dataset `weather` contains weather data for the month of January for Kalamazoo, Michigan, from 1900 to 1995. For this example, our response variable is
`avemax` which is the average maximum temperature in January. The top panel
of Figure 4.12 shows the response variable over time. There seems to be little
trend in the data. The lower panel shows the `loess` fits, local LS (solid line)
and local robust (broken line). The fits are about the same. They do show a
slight pattern of a warming trend between 1930 and 1950. ∎

4.8 Correlation

In a simple linear regression problem involving a response variable Y and a
predictor variable X, the fit of the model is of main interest. In particular,

Old Faithful Data

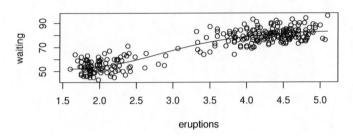

eruptions

Residual Plot

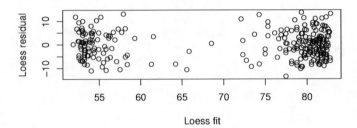

Loess fit

FIGURE 4.11
Top panel is the scatterplot of Old Faithful data, waiting time until the next eruption versus duration of eruption, overlaid with the loess fit. The bottom panel is the residual plot based on the loess fit.

we are often interested in predicting the random variable Y in terms of x and we treat x as nonstochastic. In certain settings, though, the pair (X, Y) is taken to be a random vector and we are interested in measuring the strength of a relationship or association between the random variables X and Y. By no association, we generally mean that the random variables X and Y are independent, so the basic hypotheses of interest in this section are:

$$H_0 : \ X \text{ and } Y \text{ are independent versus } H_A : X \text{ and } Y \text{ are dependent.}$$
(4.30)

For this section, assume that (X, Y) is a continuous random vector with joint cdf and pdf $F(x, y)$ and $f(x, y)$, respectively. Recall that X and Y are independent random variables if their joint cdf factors into the product of the marginal cdfs; i.e., $F(x, y) = F_X(x)F_Y(y)$ where $F_X(x)$ and $F_Y(y)$ are the marginal cdfs of X and Y, respectively. In Section 2.7 we discussed a χ^2 goodness-of-fit test for independence when X and Y are discrete random variables. For the discrete case, independence is equivalent to the statement

Ave Max Temp per Year

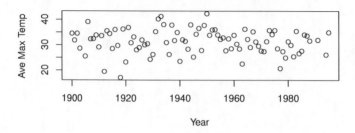

Local Ls loess (Solid) Robust (Broken)

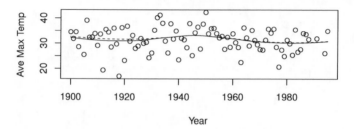

FIGURE 4.12
Top panel shows scatterplot of average maximum January temperature in
Kalamazoo, MI, over the years from 1900 to 1995. Lower panel displays the
local LS (solid line) and the local robust (broken line) loess fits.

$P(X = x, Y = y) = P(X = x)P(Y = y)$ for all x and y. In fact, the null
expected frequencies of the χ^2 goodness-of-fit test statistic are based on this
statement.[2] In the continuous case, we consider **measures of association**
between X and Y. We discuss the traditional Pearson's measure of association (the correlation coefficient ρ) and two popular nonparametric measures
(Kendall's τ_K and Spearman's ρ_S).

Let $(X_1, Y_1), (X_2, Y_2), \ldots, (X_n, Y_n)$ denote a random sample of size n on
the random vector (X, Y). Using this notation, we discuss the estimation,
associated inference, and R computation for these measures. Our discussion
is brief and many of the facts that we state are discussed in detail in most
introductory mathematical statistics texts; see, for example, Chapters 9 and
10 of Hogg et al. (2013).

---

[2]For the continuous case, this statement is simply $0 = 0$, so the χ^2 goodness-of-fit test
is not an option.

4.8.1   Pearson's Correlation Coefficient

The traditional correlation coefficient between X and Y is the ratio of the covariance between X and Y to the product of their standard deviations, i.e.,

$$\rho = \frac{E[(X - \mu_X)(Y - \mu_Y)]}{\sigma_X \sigma_Y}, \qquad (4.31)$$

where μ_X, σ_X and μ_Y, σ_Y are the respective means and standard deviations of X and Y. The parameter ρ requires, of course, the assumption of finite variance for both X and Y. It is a measure of linear association between X and Y. It can be shown that it satisfies the properties: $-1 \leq \rho \leq 1$; $\rho = \pm 1$ if and only if Y is a linear function of X (with probability 1); and $\rho > (<) 0$ is associated with a positive (negative) linear relationship between Y and X. Note that if X and Y are independent then $\rho = 0$. In general, the converse is not true. The contrapositive, though, is true; i.e., $\rho \neq 0 \Rightarrow X$ and Y are dependent.

Usually ρ is estimated by the nonparametric estimator. The numerator is estimated by the sample covariance, $n^{-1} \sum_{i=1}^{n}(X_i - \overline{X})(Y_i - \overline{Y})$, while the denominator is estimated by the product of the sample standard deviations (with n, not $n - 1$, as divisors of the sample variances). This simplifies to the **sample correlation coefficient** given by

$$r = \frac{\sum_{i=1}^{n}(X_i - \overline{X})(Y_i - \overline{Y})}{\sqrt{\sum_{i=1}^{n}(X_i - \overline{X})^2 \cdot \sum_{i=1}^{n}(Y_i - \overline{Y})^2}}. \qquad (4.32)$$

Similarly, it can be shown that r satisfies the properties: $-1 \leq r \leq 1$; $r = \pm 1$ if there is a deterministic linear relationship for the sample (X_i, Y_i); and $r > (<) 0$ is associated with a positive (negative) linear relationship between Y_i and X_i. The estimate of the correlation coefficient is directly related to simple least squares regression. Let $\hat{\sigma}_x$ and $\hat{\sigma}_y$ denote the respective sample standard deviations of X and Y. Then we have the relationship

$$r = \frac{\hat{\sigma}_x}{\hat{\sigma}_y}\hat{\beta}, \qquad (4.33)$$

where $\hat{\beta}$ is the least squares estimate of the slope in the simple regression of Y_i on X_i. It can be shown that, under the null hypothesis, $\sqrt{n}r$ is asymptotically $N(0, 1)$. Inference for ρ can be based on this asymptotic result, but usually the t-approximation discussed next is used.

If we make the much stronger assumption that the random vector (X, Y) has a bivariate normal distribution, then the estimator r is the maximum likelihood estimate (MLE) of ρ. Based on expression (4.33) and the usual t-ratio in regression, under H_0, the statistic

$$t = \frac{\sqrt{n - 2}\, r}{\sqrt{1 - r^2}} \qquad (4.34)$$

has t-distribution with $n - 2$ degrees of freedom; see page 508 of Hogg et al.

(2013). Thus a level α test of the hypotheses (4.30) is to reject H_0 in favor of H_A if $|t| > t_{\alpha/2, n-2}$. Furthermore, for general ρ, it can be shown that $\log[(1+r)/(1-r)]$ is approximately normal with mean $\log[(1+\rho)/(1-\rho)]$. Based on this, approximate confidence intervals for ρ can be constructed. In practice, usually the strong assumption of bivariate normality cannot be made. In this case, the t-test and confidence interval are approximate. For computation in R, assume that the R vectors x and y contain the samples $X_1, \ldots, X_n$ and $Y_1, \ldots, Y_n$, respectively. Then the R function cor.test computes this analysis; see Example 4.8.1 below. If inference is not needed the function cor may be used to just obtain the estimate.

4.8.2   Kendall's τ_K

Kendall's τ_K is the first nonparametric measure of association that we discuss. As above, let (X, Y) denote a jointly continuous random vector. Kendall's τ_K is a measure of **monotonicity** between X and Y. Let the two pairs of random variables (X_1, Y_1) and (X_2, Y_2) be independent random vectors with the same distribution as (X, Y). We say that the pairs (X_1, Y_1) and (X_2, Y_2) are **concordant** or **discordant** if

$$\text{sign}\{(X_1 - X_2)(Y_1 - Y_2)\} = 1 \text{ or sign}\{(X_1 - X_2)(Y_1 - Y_2)\} = -1,$$

respectively. Concordant pairs are indicative of increasing monotonicity between X and Y, while discordant pairs indicate decreasing monotonicity. Kendall's τ_K measures this monotonicity in a probability sense. It is defined by

$$\tau_K = P[\text{sign}\{(X_1 - X_2)(Y_1 - Y_2)\} = 1] - P[\text{sign}\{(X_1 - X_2)(Y_1 - Y_2)\} = -1]. \tag{4.35}$$

It can be shown that $-1 \leq \tau_K \leq 1$; $\tau_K > 0$ indicates increasing monotonicity; $\tau_K < 0$ indicates decreasing monotonicity; and $\tau_K = 0$ reflects neither monotonicity. It follows that if X and Y are independent then $\tau_K = 0$. While the converse is not true, the contrapositive is true; i.e., $\tau_K \neq 0 \Rightarrow X$ and Y are dependent.

Using the random sample $(X_1, Y_1), (X_2, Y_2), \ldots, (X_n, Y_n)$, a straightforward estimate of τ_K is simply to count the number of concordant pairs in the sample and subtract from that the number of discordant pairs. Standardization of this statistic leads to

$$\hat{\tau}_K = \binom{n}{2}^{-1} \sum_{i<j} \text{sign}\{(X_i - X_j)(Y_i - Y_j)\} \tag{4.36}$$

as our estimate of τ_K. Since the statistic $\hat{\tau}_K$ is a Kendall's τ_K based on the empirical sample distribution, it shares the same properties; i.e., $\hat{\tau}_K$ is between -1 and 1; positive values of $\hat{\tau}_K$ reflect increasing monotonicity; and negative values reflect decreasing monotonicity. It can be shown that $\hat{\tau}_K$ is an unbiased estimate of τ_K. Further, under the assumption that X and Y are independent, the statistic $\hat{\tau}_K$ is distribution-free with mean 0 and variance $2(2n + 5)/[9n(n - 1)]$. Tests of the hypotheses (4.30) can be based on the exact finite sample distribution. Tie corrections for the test are available. Furthermore, distribution-free confidence intervals[3] for τ_K exist. R computation of the inference for Kendall's τ_K is obtained by the function `cor.test` with `method="kendall"`; see Example 4.8.1. Although this R function does not compute a confidence interval for τ_K, in Section 4.8.4 we provide an R function to compute the percentile bootstrap confidence interval for τ_K.

4.8.3   Spearman's ρ_S

In defining Spearman's ρ_S, it is easier to begin with its estimator. Consider the random sample $(X_1, Y_1), (X_2, Y_2), \ldots, (X_n, Y_n)$. Denote by $R(X_i)$ the rank of X_i among $X_1, X_2, \ldots, X_n$ and likewise define $R(Y_i)$ as the rank of Y_i among $Y_1, Y_2, \ldots, Y_n$. The estimate of ρ_S is simply the sample correlation coefficient with X_i and Y_i replaced respectively by $R(X_i)$ and $R(Y_i)$. Let r_S denote this correlation coefficient. Note that the denominator of r_S is a constant and that the sample mean of the ranks is $(n + 1)/2$. Simplification leads to the formula

$$r_S = \frac{\sum_{i=1}^{n}(R(X_i) - [(n + 1)/2])(R(Y_i) - [(n + 1)/2])}{n(n^2 - 1)/12}. \tag{4.37}$$

This statistic is a correlation coefficient, so it is between ± 1, It is ± 1 if there is a strictly increasing (decreasing) relation between X_i and Y_i; hence, similar to Kendall's $\hat{\tau}_K$, it estimates monotonicity between the samples. It can be shown that

$$E(r_S) = \frac{3}{n + 1}[\tau_K + (n - 2)(2\gamma - 1)],$$

where $\gamma = P[(X_2 - X_1)(Y_3 - Y_1) > 0]$. The parameter that r_S is estimating is not as easy to interpret as the parameter τ_K.

If X and Y are independent, it follows that r_S is a distribution-free statistic with mean 0 and variance $(n - 1)^{-1}$. We accept $H_A : X$ and Y are dependent for large values of $|r_S|$. This test can be carried out using the exact distribution or approximated using the z-statistic $\sqrt{n - 1}r_S$. In applications, however, similar to expression (4.34), the t-approximation[4] is often used, where

$$t = \frac{\sqrt{n - 2}r_S}{\sqrt{1 - r_S^2}}. \tag{4.38}$$

There are distribution-free confidence intervals for ρ_S and tie corrections[5] are available. The R command `cor.test` with `method="spearman"` returns the

---

[3]See Chapter 8 of Hollander and Wolfe (1999).
[4]See, for example, page 347 of Huitema (2011).
[5]See Hollander and Wolfe (1999).

analysis based on Spearman's r_S. This computes the test statistic and the p-value, but not a confidence interval for ρ_S. Although the parameter ρ_S is difficult to interpret, nevertheless confidence intervals are important for they give a sense of the strength (effect size) of the estimate. As with Kendall's τ_K, in Section 4.8.4, we provide an R function to compute the percentile bootstrap confidence interval for ρ_S.

Remark 4.8.1 (Hypothesis Testing for Associations). In general, let ρ_G denote any of the measures of association discussed in this section. If $\rho_G \neq 0$ then X and Y are dependent. Hence, if the statistical test rejects $\rho_G = 0$, then we can statistically accept H_A that X and Y are dependent. On the other hand, if $\rho_G = 0$, then X and Y are not necessarily independent. Hence, if the test fails to accept H_A, then we should not conclude independence between X and Y. In this case, we should conclude that there is not significant evidence to refute $\rho_G = 0$. ∎

4.8.4 Computation and Examples

We illustrate the R function `cor.test` in the following example.

Example 4.8.1 (Baseball Data, 2010 Season). Datasets of major league baseball statistics can be downloaded at the site `baseballguru.com`. For this example, we investigate the relationship between the batting average of a full-time player and the number of home runs that he hits. By full-time we mean that the batter had at least 450 official at bats during the season. These data are in the `npsm` dataset `bb2010`. Figure 4.13 displays the scatterplot of home run production versus batting average for full-time players. Based on this plot there is an increasing monotone relationship between batting average and home run production, although the relationship is not very strong.

In the next code segment, the R analyses (based on `cor.test`) of Pearson's, Spearman's, and Kendall's measures of association are displayed.

```
> with(bb2010,cor.test(ave,hr))

        Pearson's product-moment correlation

data:  ave and hr
t = 2.2719, df = 120, p-value = 0.02487
alternative hypothesis: true correlation is not equal to 0
95 percent confidence interval:
 0.02625972 0.36756513
sample estimates:
      cor
0.2030727

> with(bb2010,cor.test(ave,hr,method="spearman"))
```

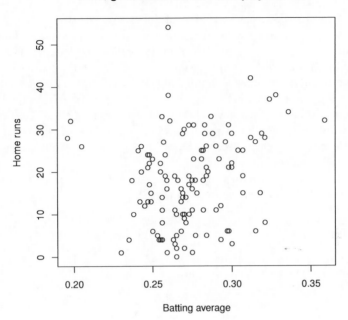

Batting statistics for full-time players, 2010

FIGURE 4.13
Scatterplot of home runs versus batting average for players who have at least
450 at bats during the 2010 Major League Baseball Season.

```
        Spearman's rank correlation rho

data:  ave and hr
S = 234500, p-value = 0.01267
alternative hypothesis: true rho is not equal to 0
sample estimates:
      rho
0.2251035

> with(bb2010,cor.test(ave,hr,method="kendall"))

        Kendall's rank correlation tau

data:  ave and hr
z = 2.5319, p-value = 0.01134
alternative hypothesis: true tau is not equal to 0
sample estimates:
```

```
        tau
0.1578534
```

For each of the methods the output contains a test statistic and associated p-value as well as the point estimate of the measure of association. Pearson's also contains the estimated confidence interval (95% by default). For example the results from Pearson's analysis give $r = 0.203$ and a p-value of 0.025. While all three methods show a significant positive association between home run production and batting average, the results for Spearman's and Kendall's procedures are somewhat stronger than that of Pearson's. Based on the scatterplot, Figure 4.13, there are several outliers in the dataset which may have impaired Pearson's r. On the other hand, Spearman's r_S and Kendall's $\hat{\tau}_K$ are robust to the effects of the outliers.

The output for Spearman's method results in the value of r_S and the p-value of the test. It also computes the statistic

$$S = \sum_{i=1}^{n} [R(X_i) - R(Y_i)]^2.$$

Although it can be shown that

$$r_S = 1 - \frac{6S}{n^2 - n};$$

the statistic S does not readily show the strength of the association, let alone the sign of the monotonicity. Hence, in addition, we advocate forming the z statistic or the t-approximation of expression (4.38). The latter gives the value of 2.53 with an approximate p-value of 0.0127. This p-value agrees with the p-value calculated by `cor.test` and the value of the standardized test statistic is readily interpreted. See Exercise 4.9.16.

As with the R output for Spearman's procedure, the output for Kendall's procedure includes $\hat{\tau}_K$ and the p-value of the associated test. The results of the analysis based on Kendall's procedure indicate that there is a significant monotone increasing relationship between batting average and home run production, similar to the results for Spearman's procedure. The estimate of association is smaller than that of Spearman's, but recall that they are estimating different parameters. Instead of a z-test statistic, R computes the test statistic T which is the number of pairs which are monotonically increasing. It is related to $\hat{\tau}_K$ by the expression

$$T = \left\{ \binom{n}{2} [1 + \hat{\tau}_K] \right\} / 2.$$

The statistic T does not lend itself easily to interpretation of the test. Even the sign of monotonicity is missing. As with the Spearman's procedure, we recommend also computing the standardized test statistic; see Exercise 4.9.17.

In general, a confidence interval yields a sense of the strength of the relationship. For example, a "quick" standard error is the length of a 95% confidence interval divided by 4. The function `cor.test` does not compute confidence intervals for Spearman's and Kendall's methods. We have written an R function, `cor.boot.ci`, which obtains a percentile bootstrap confidence interval for each of the three measures of association discussed in this section. Let $B = [X, Y]$ be the matrix with the samples of X_i's in the first column and the samples of Y_i's in the second column. Then the bootstrap scheme resamples the rows of B with replacement to obtain a bootstrap sample of size n. This is performed n_{BS} times. For each bootstrap sample, the estimate of the measure of association is obtained. These bootstrap estimates are collected and the $\alpha/2$ and $(1 - \alpha/2)$ percentiles of this collection form the confidence interval. The default arguments of the function are:

```
> args(cor.boot.ci)

function (x, y, method = "spearman", conf = 0.95, nbs = 3000)
NULL
```

Besides Spearman's procedure, bootstrap percentile confidence intervals are computed for ρ and τ_K by using respectively the arguments `method="pearson"` and `method="kendall"`. Note that $(1 - \alpha)$ is the confidence level and the default number of bootstrap samples is set at 3000. We illustrate this function in the next example.

Example 4.8.2 (Continuation of Example 4.8.1). The code segment below obtains a 95% percentile bootstrap confidence interval for Spearman's ρ_S.

```
> library(boot)
> with(bb2010,cor.boot.ci(ave,hr))

      2.5%       97.5%
0.05020961 0.39888150
```

The following code segment computes percentile bootstrap confidence intervals for Pearson's and Kendall's methods.

```
> with(bb2010,cor.boot.ci(ave,hr,method='pearson'))

       2.5%        97.5%
0.005060283 0.400104126

> with(bb2010,cor.boot.ci(ave,hr,method='kendall'))

      2.5%       97.5%
0.02816001 0.28729659
```

TABLE 4.1

Estimates and Confidence Intervals for the Three Methods.
The first three columns contain the results for the original
data, while the last three columns contain the results for the
changed data.

	Original Data			Outlier Data		
	Est	LBCI	UBCI	Est2	LBCI2	UBCI2
Pearson's	0.20	0.00	0.40	0.11	0.04	0.36
Spearman's	0.23	0.04	0.40	0.23	0.05	0.41
Kendall's	0.16	0.03	0.29	0.16	0.04	0.29

To show the robustness of Spearman's and Kendall's procedures, we changed the home run production of the 87*th* batter from 32 to 320; i.e., a typographical error. Table 4.1 compares the results for all three procedures on the original and changed data.[6]

Note that the results for Spearman's and Kendall's procedures are essentially the same on the original dataset and the dataset with the outlier. For Pearson's procedure, though, the estimate changes from 0.20 to 0.11. Also, the confidence interval has been affected. ∎

4.9   Exercises

4.9.1. Obtain a scatterplot of the `telephone` data. Overlay the least squares and R fits.

4.9.2. Write an R function which given the results of a call to `rfit` returns the diagnostic plots: Studentized residuals versus fitted values, with ±2 horizontal lines for outlier identification; normal $q-q$ plot of the Studentized residuals, with ±2 horizontal lines outliers for outlier identification; histogram of residuals; and a boxplot of the residuals.

4.9.3. Consider the free fatty acid data.

(a) For the Wilcoxon fit, obtain the Studentized residual plot and $q-q$ plot of the Studentized residuals. Comment on the skewness of the errors.

(b) Redo the analysis of the free fatty acid data using the bent scores (`bentscores1`). Compare the summary of the regression coefficients with those from the Wilcoxon fit. Why is the bent score fit more precise (smaller standard errors) than the Wilcoxon fit?

---

[6]These analyses were run in a separate step so they may differ slightly from those already reported.

4.9.4. Using the `baseball` data, calculate a large sample confidence interval for the slope parameter when regressing weight on height. Compare the results to those obtained using the bootstrap discussed in Section 4.6.

4.9.5. Consider the following data:

x	1	2	3	4	5	6	7	8	9	10	11	12	13	14	15
y	−7	0	5	9	−3	−6	18	8	−9	−20	−11	4	−1	7	5

Consider the simple regression model: $Y = \beta_0 + \beta_1 x + e$.

(a) For Wilcoxon scores, write R code which obtains a sensitivity curve of the `rfit` of the estimate of β_1, where the sensitivity curve is the difference in the estimates of β_1 between perturbed data and the original data.

(b) For this exercise, use the above data as the original data. Let $\hat{\beta}_1$ denote the Wilcoxon estimate of slope based on the original data. Then obtain 9 perturbed datasets using the following sequence of replacements to y_{15}: $−995, −95, −25, −5, 5, 10, 30, 100, 1000$. Let $\hat{\beta}_{1j}$ be the Wilcoxon fit of the jth perturbed dataset for $j = 1, 2, \ldots, 9$. Obtain the sensitivity curve which is the plot of $\hat{\beta}_{1j} − \hat{\beta}_1$ versus the jth replacement value for y_{15}.

(c) Obtain the sensitivity curve for the LS fit. Compare it with the Wilcoxon sensitivity curve.

4.9.6. For the simple regression model, the estimator of slope proposed by Theil (1950) is defined as the median of the pairwise slopes:

$$\hat{\beta}_T = \text{med}\{b_{ij}\}$$

where $b_{ij} = (y_j − y_i)/(x_j − x_i)$ for $i < j$.

(a) Write an R function which takes as input a vector of response variables and a vector of explanatory variables and returns the Theil estimate.

(b) For a simple regression model where the predictor is a continuous variable, write an R function which computes the bootstrap percentile confidence interval for the slope parameter based on Theil's estimate.

(c) Show that Theil's estimate reduces to the the Hodges–Lehmann estimator for the two-sample location problem.

4.9.7. Consider the data of Example 4.4.1.

(a) Obtain a scatterplot of the data.

(b) Obtain the Wilcoxon fit of the linear trend model. Overlay the fit on the scatterplot. Obtain the Studentized residual plot and normal $q-q$ plots. Identify any outliers and comment on the quality of fit.

(c) Obtain a 95% confidence interval for the slope parameter and use it to test the hypothesis of 0 slope.

(d) Estimate the mean of the response when time has the value 16 and find the 95% confidence interval for it which was discussed in Section 4.4.4.

4.9.8. Bowerman et al. (2005) present a dataset concerning the value of a home (x) and the upkeep expenditure (y). The data are in `qhic`. The variable x is in \$1000's of dollars while the y variable is in \$10's of dollars.

(a) Obtain a scatterplot of the data.

(b) Use Wilcoxon Studentized residual plots, values of $\hat{\tau}$, and values of the robust R^2 to decide whether a linear or a quadratic model fits the data better.

(c) Based on your model, estimate the expected expenditures (with a 95% confidence interval) for a house that is worth \$155,000.

(d) Repeat (c) for a house worth \$250,000.

4.9.9. Rewrite the `aligned.test` function to take an additional design matrix as its third argument instead of group/treatment membership. That is, for the model $Y = \alpha 1 + X_1\beta_1 + X_2\beta_2 + e$, test the hypothesis $H_0 : \beta_2 = 0$.

4.9.10. Hettmansperger and McKean (2011) discuss a dataset in which the dependent variable is the cloud point of a liquid, a measure of degree of crystallization in a stock, and the independent variable is the percentage of I-8 in the base stock. For the readers' convenience, the data can be found in the dataset `cloud` in the package `npsm`.

(a) Scatterplot the data. Based on the plot, is a simple linear regression model appropriate?

(b) Show by residual plots of the fits that the linear and quadratic polynomials are not appropriate but that the cubic model is.

(c) Use the R function `polydeg`, with a super degree set at 5, to determine the degree of the polynomial. Compare with Part (b).

4.9.11. Devore (2012) discusses a dataset on energy. The response variable is the energy output in watts while the independent variable is the temperature difference in degrees K. A polynomial fit is suggested. The data are in the dataset `energy`.

(a) Scatterplot the data. What degree of polynomial seems suitable?

(b) Use the R function `polydeg`, with a super degree set at 6, to determine the degree of the polynomial.

(c) Based on a residual analysis, does the polynomial fit of Part (b) provide a good fit?

4.9.12. Consider the weather dataset, `weather`, discussed in Example 4.7.4. One of the variables is mean average temperature for the month of January (`meantmp`).

(a) Obtain a scatterplot of the mean average temperature versus the year. Determine the warmest and coldest years.

(b) Obtain the `loess` fit of the data. Discuss the fit in terms of years, (were there warm trends, cold trends?).

4.9.13. As in the last problem, consider the weather dataset, `weather`. One of the variables is total snowfall (in inches), `totalsnow`, for the month of January.

(a) Scatterplot total snowfall versus year. Determine the years of maximal and minimal snowfalls.

(b) Obtain the local LS and robust `loess` fits of the data. Compare the fits.

(c) Perform a residual analysis on the robust fit.

(d) Obtain a boxplot of the residuals found in Part (c). Identify the outliers by year.

4.9.14. In the discussion of Figure 4.7, the nonparametric regression fit by `ksmooth` detects an artificial valley. Obtain the locally robust loess fit of this dataset (`poly`) and compare it with the `ksmooth` fit.

4.9.15. Using the `baseball` data, obtain the scatterplot between the variables home run productions and RBIs. Then compute the Pearson's, Spearman's, and Kendall's analyses for these variables. Comment on the plot and analyses.

4.9.16. Write an R function which computes the t-test version of Spearman's procedure and returns it along with the corresponding p-value and the estimate of ρ_S.

4.9.17. Repeat Exercise 4.9.16 for Kendall's procedure.

4.9.18. Create a graphic similar to Figure 4.10.

4.9.19. Recall that, in general, the three measures of association estimate different parameters. Consider bivariate data (X_i, Y_i) generated as follows:

$$Y_i = X_i + e_i, \quad i = 1, 2, \ldots, n,$$

where X_i has a standard Laplace (double exponential) distribution, e_i has a standard $N(0, 1)$ distribution, and X_i and e_i are independent.

(a) Write an R script which generates this bivariate data. The supplied R function `rlaplace(n)` generates n iid Laplace variates. For $n = 30$, compute such a bivariate sample. Then obtain the scatterplot and the association analyses based on the Pearson's, Spearman's, and Kendall's procedures.

(b) Next write an R script which simulates the generation of these bivariate samples and collects the three estimates of association. Run this script for 10,000 simulations and obtain the sample averages of these estimates, their corresponding standard errors, and approximate 95% confidence intervals. Comment on the results.

4.9.20. The electronic memory game Simon was first introduced in the late 1970s. In the game there are four colored buttons which light up and produce a musical note. The device plays a sequence of light/note combinations and the goal is to play the sequence back by pressing the buttons. The game starts with one light/note and progressively adds one each time the player correctly recalls the sequence.[7]

Suppose the game were played by a set of statistics students in two classes (time slots). Each student played the game twice and recorded his or her longest sequence. The results are in the dataset `simon`.

Regression toward the mean is the phenomenon that if an observation is extreme on the first trial it will be closer to the average on the second trial. In other words, students that scored higher than average on the first trial would tend to score lower on the second trial and students who scored low on the first trial would tend to score higher on the second.

(a) Obtain a scatterplot of the data.

(b) Overlay an R fit of the data. Use Wilcoxon scores. Also overlay the line $y = x$.

(c) Obtain an R estimate of the slope of the regression line as well as an associated confidence interval.

(d) Do these data suggest a regression toward the mean effect?

---

[7]The game is implemented on the web. The reader is encouraged to use his or her favorite search engine and try it out.

5

ANOVA and ANCOVA

5.1 Introduction

In this chapter, the R functions in the packages `Rfit` and `npsm` for the computation of fits and inference for standard rank-based analyses of variance (ANOVA)[1] and analysis of covariance (ANCOVA) type designs are discussed. These include one-way, two-way, and k-way crossed designs that are covered in Sections 5.2–5.4. Both tests of general linear hypotheses and estimation of effects with standard errors and confidence intervals are emphasized. We also briefly present multiple comparison procedures (MCP), in particular a robust Tukey–Kramer procedure, illustrating their computation via the package `Rfit` . We also consider the R computation of several traditional nonparametric methods for these designs including the Kruskal–Wallis (Section 5.2.2) and the Jonckheere–Terpstra tests for ordered alternatives (Section 5.6). In the last section, the generalization of the Fligner–Kileen procedure of Chapter 3 to the k-sample scale problem is presented. The rank-based analyses covered in this chapter are for fixed effect models. Ranked-based methods and their computation for mixed (fixed and random) models form the topic of Chapter 8.

As a cursory reading, we suggest Section 5.2 and the two-way design material of Section 5.3, and the ordered alternative methods of Section 5.6. As usual, our emphasis is on how to easily compute these rank-based procedures using `Rfit`. Details of the robust rank-based inference for these fixed effect models are discussed in Chapter 4 of Hettmansperger and McKean (2011).

5.2 One-Way ANOVA

Suppose we want to determine the effect that a single factor A has on a response of interest over a specified population. Assume that A consists of k levels or treatments. In a completely randomized design (CRD), n subjects are randomly selected from the reference population and n_i of them are ran-

---

[1]Though could be named ANODI for rank-based analysis.

domly assigned to level i, $i = 1, \ldots k$. Let the jth response in the ith level be denoted by Y_{ij}, $j = 1, \ldots, n_i$, $i = 1, \ldots, k$. We assume that the responses are independent of one another and that the distributions among levels differ by at most shifts in location.

Under these assumptions, the **full model** can be written as

$$Y_{ij} = \mu_i + e_{ij} \quad j = 1, \ldots, n_i \, , \, i = 1, \ldots, k \, , \tag{5.1}$$

where the e_{ij}s are iid random variables with density $f(x)$ and distribution function $F(x)$ and the parameter μ_i is a convenient location parameter for the ith level, (for example, the mean or median of the ith level). This model is often referred to as a one-way design and its analysis as a one-way analysis of variance (ANOVA). Generally, the parameters of interest are the effects (pairwise contrasts),

$$\Delta_{ii'} = \mu_{i'} - \mu_i, \quad i \neq i', 1, \ldots, k. \tag{5.2}$$

We can express the model in terms of these simple contrasts. As in the R `lm` command, we reference the first level. Then the Model (5.1) can be expressed as

$$Y_{ij} = \begin{cases} \mu_1 + e_{1j} & j = 1, \ldots, n_1 \\ \mu_1 + \Delta_{i1} + e_{ij} & j = 1, \ldots, n_i, \, i = 2, \ldots, k. \end{cases} \tag{5.3}$$

Let $\mathbf{\Delta} = (\Delta_{21}, \Delta_{31}, \ldots, \Delta_{k1})'$. Upon fitting the model a residual analysis should be conducted to check these model assumptions. As the full model fit is based on a linear model, the diagnostic procedures discussed in Chapter 4 are implemented for ANOVA and ANCOVA models as well.

Observational studies can also be modeled this way. Suppose k independent samples are drawn, one from each of k populations. If we assume further that the distributions of the populations differ by at most a shift in locations then Model (5.1) is appropriate. Usually, in the case of observational studies, it is necessary to adjust for covariates. These analyses are referred to as the analysis of covariance and are discussed in Section 5.4.

The analysis for the one-way design is usually a test of the hypothesis that all the effects, Δ_i's, are 0, followed by individual comparisons of levels. The hypothesis can be written as

$$H_0 : \mu_1 = \cdots = \mu_k \text{ versus } H_A : \mu_i \neq \mu_{i'} \text{ for some } i \neq i'. \tag{5.4}$$

Confidence intervals for the simple contrasts $\Delta_{ii'}$ can be used for the pairwise comparisons. We next briefly describe the general analysis for the one-way model and discuss its computation by `Rfit`.

A test of the overall hypothesis (5.4) is based on a reduction in dispersion test, first introduced in (4.4.3). For `Rfit`, assume that a score function φ has been selected; otherwise, `Rfit` uses the default Wilcoxon score function. As discussed in Section 5.3, let $\widehat{\mathbf{\Delta}}_\varphi$ be the rank-based estimate of $\mathbf{\Delta}$ when the full model, (5.1), is fit. Let $D_\varphi(\text{FULL}) = D(\widehat{\mathbf{\Delta}}_\varphi)$ denote the full model dispersion, i.e., the minimum value of the dispersion function when this full model is fit. The reduced model is the location model

$$Y_{ij} = \mu + e_{ij} \quad j = 1, \ldots, n_i \,, \; i = 1, \ldots, k. \tag{5.5}$$

Because the dispersion function is invariant to location, the minimum dispersion at the reduced model is the dispersion of the observations; i.e., $D(\mathbf{0})$ which we call $D_\varphi(\text{RED})$. The reduction in dispersion is then $RD_\varphi = D_\varphi(\text{RED}) - D_\varphi(\text{FULL})$ and, hence, the drop in dispersion test statistic is given by

$$F_\varphi = \frac{RD_\varphi/(k-1)}{\widehat{\tau}_\varphi/2}, \tag{5.6}$$

where $\widehat{\tau}_\varphi$ is the estimate of scale discussed in Section 3.1. The approximate level α test rejects H_0, if $F_\varphi \geq F_{\alpha, k-1, n-k}$. The traditional LS test is based on a reduction of sums of squares. Replacing this by a reduction in dispersion the test based on F_φ can be summarized in an ANOVA table much like that for the traditional F-test; see page 298 of Hettmansperger and McKean (2011). When the linear Wilcoxon scores are used, we often replace the subscript φ by the subscript W; that is, we write the Wilcoxon rank-based F-test statistic as

$$F_W = \frac{RD_W/(k-1)}{\widehat{\tau}_W/2}. \tag{5.7}$$

The Rfit function `oneway.rfit` computes the robust rank-based one-way analysis. Its arguments are the vector of responses and the corresponding vector of levels. It returns the value of the test statistic and the associated p-value. We illustrate its computation with an example.

Example 5.2.1 (LDL Cholesterol of Quail). Hettmansperger and McKean (2011), page 295, discuss a study which investigated the effect that four drug compounds had on the reduction of low density lipid (LDL) cholesterol in quail. The drug compounds are labeled as I, II, III, and IV. The sample size for each of the first three levels is 10 while 9 quail received compound IV. The boxplots shown in Figure 5.1 attest to a difference in the LDL levels over treatments.

Using Wilcoxon scores, the results of `oneway.rfit` are:

```
> robfit = with(quail,oneway.rfit(ldl,treat))
> robfit

Call:
oneway.rfit(y = ldl, g = treat)

Overall Test of All Locations Equal

Drop in Dispersion Test
F-Statistic      p-value
```

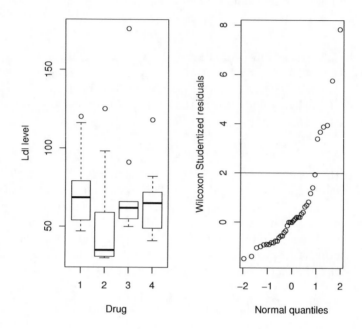

FIGURE 5.1
Plots for LDL cholesterol of quail example.

```
    3.916371       0.016404


              Pairwise comparisons using Rfit

data:  ldl and treat

    2     3    4
1 -     -    -
2 1.00 -    -
3 0.68 0.99 -
4 0.72 0.99 0.55

P value adjustment method: none
```

The Wilcoxon test statistic has the value $F_W = 3.92$ with p-value 0.0164. Thus the Wilcoxon test indicates that the drugs differ in their lowering of cholesterol effect. In contrast to the highly significant Wilcoxon test of the hypothesis (5.4), the LS-based F-test statistic has value 1.14 with the p-value 0.3451. In practice, using the LS results, one would not proceed with comparisons of the drugs with such a large p-value. Thus, for this dataset, the robust and LS analyses would have different practical interpretations. Also, the coefficient of precision, (3.46), for this data between the Wilcoxon and LS analyses is $\widehat{\sigma}^2/\widehat{\tau}^2 = 2.72$. Hence, the Wilcoxon analysis is much more precise.

The resulting $q-q$ plot (see right panel of Figure 5.1) of the Studentized Wilcoxon residuals indicates that the random errors e_{ij} have a skewed distribution. R fits based on scores more appropriate than the Wilcoxon for skewed errors are discussed later.

5.2.1 Multiple Comparisons

The second stage of an analysis of a one-way design consists of pairwise comparisons of the treatments. The robust $(1 - \alpha)100\%$ confidence interval to compare the ith and ith' treatments is given by

$$\widehat{\Delta}_{ii'} \pm t_{\alpha/2,n-1}\widehat{\tau}\sqrt{\frac{1}{n_i} + \frac{1}{n_{i'}}}. \tag{5.8}$$

Often there are many comparisons of interest. For example, in the case of all pairwise comparisons $\binom{k}{2}$ confidence intervals are required. Hence, the overall family error rate is usually of concern. Multiple comparison procedures (MCP) try to control the overall error rate to some degree. There are many robust versions of MCPs from which to choose. The `summary` function associated with `oneway.rfit` computes three of the most popular of such procedures. Assuming that the fit of the full model is in `robfit`, the syntax of the command is `summary(robfit,method="none")`. The argument of `method` produces the following MCPs:

`method="none"`	No adjustment made
`method="tukey"`	Tukey–Kramer
`method="bonferroni"`	Bonferroni

We give a brief description of these procedures, followed by an example using `Rfit` .

A protected least significant difference procedure (PLSD) consists of testing the hypothesis (5.4) at a specified level α. If H_0 is rejected then the comparisons are based on the confidence intervals (5.8) with confidence coefficient $1 - \alpha$. On the other hand, if H_0 is not rejected then the analysis stops. Although this procedure does not control the overall family rate of error, the initial F-test offers protection, which has been confirmed (See, for example, McKean et al. (1989)) in simulation studies.

For the Tukey–Kramer procedure, Studentized range critical values replace the t-critical values in the intervals (5.8). When the traditional procedure is used, the random errors have a normal distribution, and the design is balanced then the Tukey–Kramer procedure has family error rate α. When either of these assumptions fail then the Tukey–Kramer procedure has a family error rate approximately equal to α.

The Bonferroni procedure depends on the number of comparisons made. Suppose there are l comparisons of interest, then the Bonferroni procedure uses the intervals (5.8) with the critical value $t_{\alpha/(2l),n-k}$. The Bonferroni procedure has overall family rate $\leq \alpha$. If all pairwise comparisons are desired then $l = \binom{k}{2}$.

Example 5.2.2 (LDL Cholesterol of Quail, Continued). For the quail data, we selected the Tukey–Kramer procedure for all six pairwise comparisons. The Rfit computation is:

```
> summary(robfit,method="tukey")
```

```
Multiple Comparisons
Method Used   tukey
```

I J	Estimate	St Err	Lower Bound CI	Upper Bound CI
1 1 2	-25.00720	8.26820	-47.30572	-2.70868
2 1 3	-3.99983	8.26820	-26.29835	18.29869
3 1 4	-5.00027	8.49476	-27.90982	17.90928
4 2 3	21.00737	8.26820	-1.29115	43.30589
5 2 4	20.00693	8.49476	-2.90262	42.91648
6 3 4	-1.00044	8.49476	-23.91000	21.90911

The Tukey–Kramer procedure declares that the Drug Compounds I and II are statistically significantly different. ∎

5.2.2   Kruskal–Wallis Test

Note that Model (5.1) generalizes the two-sample problem of Chapter 2 to k-samples. In this section, we discuss the Kruskal–Wallis test of the hypotheses (5.4) which is a generalization of the two-sample MWW test.

Assume then that Model (5.1) holds for the responses Y_{ij}, $j = 1,\ldots n_i$ and $i = 1,\ldots,k$. As before let $n = \sum_{i=1}^{n} n_i$ denote the total sample size. Let R_{ij} denote the rank of the response Y_{ij} among all n observations; i.e., the ranking is done without knowledge of treatment. Let $\overline{R}_i$ denote the average of the ranks for sample i. The test statistic H is a standardized weighted average of the squared deviations of the $\overline{R}_i$ from the average of all ranks $(n+1)/2$. The test statistic is

$$H = \frac{12}{n(n+1)} \sum_{i=1}^{n} n_i \left(\overline{R}_i - \frac{n+1}{2} \right)^2. \tag{5.9}$$

The statistic H is the **Kruskal–Wallis** test statistic; see Kruskal and Wallis (1952). Under H_0 H is distribution-free and there are some tables available for its exact distribution.[2] It also has an asymptotic χ^2-distribution with $k-1$

---

[2]See Chapter 6 of Hollander and Wolfe (1999).

degrees of freedom under H_0. The R command is `kruskal.test`. Assume the responses are in one vector x, and the group or treatment assignments are vector g, then the call is `kruskal.test(x,g)`. In addition a `formula` can be used as in `kruskal.test(x~g)`. We illustrate the computation with the following example.

Example 5.2.3 (Mucociliary Efficiency). Hollander and Wolfe (1999), page 192, discuss a small study which assessed the mucociliary efficiency from the rate of dust in the three groups: normal subjects, subjects with obstructive airway disease, and subjects with asbestosis. The responses are the mucociliary clearance half-times of the subjects. The sample sizes are small: $n_1 = n_3 = 5$ and $n_2 = 4$. Hence, $n = 14$. The data are given in the R vectors `normal`, `obstruct`, and `asbestosis` in the following code segment which computes the Kruskal–Wallis test.

```
> normal <- c(2.9,3.0,2.5,2.6,3.2)
> obstruct <- c(3.8,2.7,4.0,2.4)
> asbestosis <- c(2.8,3.4,3.7,2.2,2.0)
> x <- c(normal,obstruct,asbestosis)
> g  <- c(rep(1,5),rep(2,4),rep(3,5))
> kruskal.test(x,g)

        Kruskal-Wallis rank sum test

data:  x and g
Kruskal-Wallis chi-squared = 0.7714, df = 2, p-value = 0.68
```

Based on this p-value, there do not appear to be differences among the groups for mucociliary efficiency. ■

Corrections for ties for the Kruskal–Wallis test are discussed in Hollander and Wolfe (1999) and `kruskal.test` does make such adjustments in its calculation. As discussed in Hettmansperger and McKean (2011), the Kruskal–Wallis test is asymptotically equivalent to the drop in dispersion test (5.6) using Wilcoxon scores.[3]

5.3   Multi-Way Crossed Factorial Design

For a Multi-way or k-way crossed factorial experimental design, the `Rfit` function `raov` computes the rank-based ANOVA for all $2^k - 1$ hypotheses, including

---

[3]This equivalence extends to local alternatives; hence, the Kruskal–Wallis and the drop in dispersion tests have the same asymptotic efficiency. In particular, for normal errors the relative efficiency of these tests relative to the traditional LS test is 0.955.

the main effects and interactions of all orders. These are the hypotheses in a standard ANOVA table for a k-way design. The design may be balanced or unbalanced. For simplicity, we briefly discuss the analysis in terms of a cell mean (median) model; see Hocking (1985) for details on the traditional LS analysis and Chapter 4 of Hettmansperger and McKean (2011) for the rank-based analysis. For clarity, we first discuss a two-way crossed factorial design before presenting the k-way design.

5.3.1   Two-Way

Let A and B denote the two factors in a two-way design with levels a and b, respectively. Let Y_{ijk} define the response for the kth replication at levels i and j of factors A and B, respectively. Then the full model can be expressed as

$$Y_{ijk} = \mu_{ij} + e_{ijk}, \quad k = 1, \ldots, n_{ij}, i = 1, \ldots, a; j = 1, \ldots, b, \quad (5.10)$$

where e_{ijk} are iid random variables with pdf $f(t)$. Since the effects of interest are contrasts in the μ_{ij}'s, these parameters can be either cell means or medians, (actually any location functional suffices). We refer to this model as the **cell mean** or **cell median** model. We assume that all $n_{ij} \geq 1$ and at least one n_{ij} exceeds 1. Let $n = \sum \sum n_{ij}$ denote the total sample size.

For the two-way model, the three hypotheses of interest are the main effects hypotheses and the interaction hypothesis.[4] For the two-way model (5.10) these hypotheses are:

$$H_{0A} : \overline{\mu}_{1.} = \cdots = \overline{\mu}_{a.} \quad \text{vs.} \quad H_{1A} : \overline{\mu}_{i.} \neq \overline{\mu}_{i'.}, \text{for } i \neq i' \quad (5.11)$$

$$H_{0B} : \overline{\mu}_{.1} = \cdots = \overline{\mu}_{.b} \quad \text{vs.} \quad H_{1B} : \overline{\mu}_{.j} \neq \overline{\mu}_{.j'}, \text{for } j \neq j' \quad (5.12)$$

$$H_{0AB} : \gamma_{11} = \gamma_{12} \cdots = \gamma_{ab} \quad \text{vs.} \quad H_{1AB} : \gamma_{ij} \neq \gamma_{i'j'}, (i,j) \neq (i',j') \quad (5.13)$$

where γ_{ij} are the interaction parameters

$$\gamma_{ij} = \mu_{ij} - \overline{\mu}_{i.} - \overline{\mu}_{.j} + \overline{\mu}_{..}.$$

$\overline{\mu}_{i.} = \frac{1}{b} \sum_{j=1}^{b} \mu_{ij}$, $\overline{\mu}_{.j} = \frac{1}{a} \sum_{j=1}^{a} \mu_{ij}$, and $\overline{\mu}_{.j} = \frac{1}{ab} \sum_{i=1}^{b} \sum_{j=1}^{a} \mu_{ij}$. The function `raov` computes the drop in dispersion tests for each of these hypotheses as illustrated in the next example.

Example 5.3.1 (Serum LH Data). Hollander and Wolfe (1999) discuss a 2×5 factorial design for a study to determine the effect of light on the release of luteinizing hormone (LH). The factors in the design are: light regimes at two levels (constant light and 14 hours of light followed by 10 hours of darkness) and a luteinizing release factor (LRF) at 5 different dosage levels. The response is the level of luteinizing hormone (LH), nanograms per ml of serum in blood samples. Sixty rats were put on test under these 10 treatment combinations,

---

[4]We have chosen Type III hypotheses which are easy to interpret even for severely unbalanced designs. Details of how `Rfit` performs this procedure are given in Section 5.5.

six rats per combination. The data are in the dataset serumLH. We first obtain
the robust rank-based ANOVA for these data. The command is raov and its
syntax is shown in the output which follows.

```
> raov(serum~light+lrfdose+light*lrfdose,data = serumLH)
```

```
Robust ANOVA Table
                DF      RD    Mean RD        F       p-value
light            1 1642.4580 1642.4580 58.397963 5.976252e-10
lrfdose          4 3027.6748  756.9187 26.912415 6.150414e-12
light:lrfdose    4  451.4822  112.8706  4.013138 6.710855e-03
```

We also obtained the LS-based ANOVA.

```
> summary(aov(serum~light+lrfdose+light*lrfdose,data = serumLH))
```

```
              Df Sum Sq Mean Sq F value   Pr(>F)
light          1 242189  242189  40.223 6.41e-08 ***
lrfdose        4 545549  136387  22.652 1.02e-10 ***
light:lrfdose  4  55099   13775   2.288   0.0729 .
Residuals     50 301055    6021
---
Signif. codes:  0 '***' 0.001 '**' 0.01 '*' 0.05 '.' 0.1 ' ' 1
```

Note that the analyses differ critically. The robust ANOVA clearly detects
the interaction with p-values less that 0.01; while, interaction is not significant
at a 5% level for the LS analysis, the p-value NA For both analyses, the average
main effects are highly significant. If the level of significance was set at 0.05,
then the two analyses would lead to different practical interpretations.

Next we examine the residuals by plotting them in a normal $q-q$ plot;
these are based on the R fit of the full model using Wilcoxon scores. The
normal $q-q$ plot of this fit is shown in the left panel of Figure 5.2. It is clear
from this plot that the errors have much heavier tails than those of a normal
distribution.

The empirical measure of precision (3.46) is $\hat{\sigma}^2/\hat{\tau}_W^2 = 1.9$. Based on the
$q-q$ plot this large value is not surprising. The right panel in Figure 5.2 is
an interaction plot of profiles based on a Wilcoxon fit. That is, in this case,
we have plotted the profiles for the light regimes over the LRF levels, using
the cell estimates based on the Wilcoxon fit. It is clear from this plot that
interaction of the factors is present, which agrees with the robust test for
interaction. The corresponding profile plot (not shown) based on cell sample
means is similar, although the outliers caused some distortion. ∎

5.3.2 k-Way

Consider a general k-way factorial design with factors $1, 2, \ldots, k$ having the
levels $l_1, l_2, \ldots, l_k$, respectively. Let $Y_{i_1 i_2 \cdots i_k, j}$ be the jth response at levels

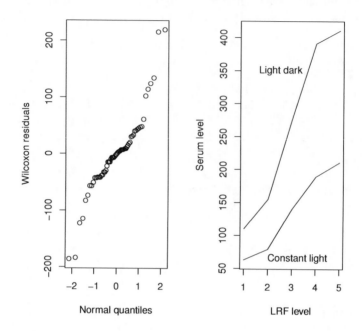

FIGURE 5.2
Plots for analysis of serum data.

$i_1 i_2 \cdots i_k$ of factors $1, 2, \ldots, k$ where $1 \leq i_1 \leq l_1$, $1 \leq i_2 \leq l_2$, $\ldots$, and $1 \leq i_k \leq l_k$ and $1 \leq j \leq n_{i_1 i_2 \cdots i_k}$. Let $n = \sum n_{i_1 i_2 \cdots i_k}$ denote the total sample size. Choose a location functional μ (can be the mean or median). Then the full model is

$$y_{i_1 i_2 \cdots i_k, j} = \mu_{i_1 i_2 \cdots i_k} + e_{i_1 i_2 \cdots i_k, j}, \qquad (5.14)$$

where the errors $e_{i_1 i_2 \cdots i_k, j}$ are iid with cdf $F(t)$ and pdf $f(t)$.

For this model there are $2^k - 1$ hypotheses of interest. These include the k main effect hypotheses and the interaction hypotheses of all orders. Using the algorithms found in Hocking (1985), the corresponding hypotheses matrices are obtained and, hence, the appropriate reduced model design matrices can be computed. An illustrative example follows.

Example 5.3.2 (Plank Data). Abebe et al. (2001) discuss a dataset resulting from a three-way layout for a neurological experiment in which the time required for a mouse to exit a narrow elevated wooden plank is measured. The response is the log of time to exit. Interest lies in assessing the effects of three factors: the Mouse Strain (Tg+, Tg-), the mouse's Gender (female, male), and the mouse's Age (Aged, Middle, Young). The design is a $2 \times 2 \times 3$ factorial design. The data are in the npsm dataset plank.

Using raov, we computed the tests for main effects, as well as the second order and third order interactions. For comparison, we also obtained the corresponding LS-based tests.

```
> raov(response~strain:gender:age, data = plank)
```

Robust ANOVA Table

	DF	RD	Mean RD	F	p-value
strain	1	4.4220740	4.4220740	13.7693442	0.0005040186
gender	1	1.9841358	1.9841358	6.1781529	0.0161864481
age	2	0.8305464	0.4152732	1.2930674	0.2831040113
strain:gender	1	0.5014476	0.5014476	1.5613950	0.2170562268
strain:age	2	1.3058129	0.6529064	2.0330038	0.1412200992
gender:age	2	2.3842111	1.1921055	3.7119486	0.0311241534
strain:gender:age	2	0.3592335	0.1796167	0.5592861	0.5750170259

```
> fit.ls = lm(response~strain:gender:age, data = plank)
> anova(fit.ls)
```

Analysis of Variance Table

Response: response

	Df	Sum Sq	Mean Sq	F value	Pr(>F)	
strain:gender:age	11	28.431	2.5847	2.1414	0.03311	*
Residuals	52	62.765	1.2070			

Signif. codes: 0 '***' 0.001 '**' 0.01 '*' 0.05 '.' 0.1 ' ' 1

If the nominal level is $\alpha = 0.05$, then, based on the rank-based ANOVA, the factors strain and gender are significant, but, so is the factor age because of its significant two-way interaction with gender. In contrast, based on the LS ANOVA, only the factor strain is significant. Thus practical interpretations would differ for the two analyses.

For diagnostic checks of the model, the Rfit function raov returns the Rfit of the full model in the value fit. Using this value, we computed the Wilcoxon Studentized residuals. Figure 5.3 displays the normal $q-q$ plot and the residual plot based on these residuals. The $q-q$ plot suggests an error distribution with very thick tails. Other than the outliers, the residual plot reveals no serious lack of fit in the model.

The difference in the Wilcoxon and LS analyses are due to these outliers. Our empirical measure of efficiency is $\hat{\sigma}^2/\hat{\tau}_W^2 = 2.93$, which indicates that the Wilcoxon analysis is almost 3 times more efficient than the LS analysis on this dataset. ∎

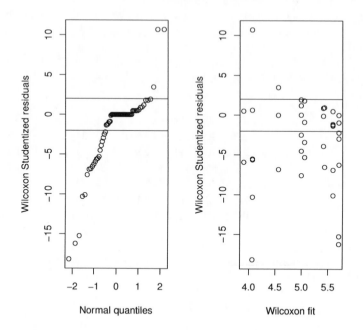

FIGURE 5.3
Diagnostic plots for plank data analysis.

5.4   ANCOVA*

It is common to collect additional information (e.g. demographics) in an experiment or study on the experimental units. When this information is included in the model to test for a treatment or group effect it can be considered an **analysis of covariance** (ANCOVA). The goal of these covariates is to account for variation in the response variables. These analyses are often called **adjusted analyses**. Absent *a priori* information of which covariates are important, a model selection procedure can be implemented with the response on the covariates; i.e., treatment is included in the model *after* the covariates are selected. In this section, we discuss this adjusted analysis for rank-based procedures. Covariate variables are sometimes referred to as concomitant variables. The monograph by Huitema (2011) offers an informative discussion of the traditional ANCOVA, while Chapter 4 of Hettmansperger and McKean (2011) presents the rank-based ANCOVA which we discuss.

For notation, consider a one-way experimental design with k different treatment combinations with cell sample sizes n_i, $i = 1, \ldots k$. The discussion easily generalizes to other ANOVA designs. Suppose there are m covariate variables. Let Y_{ij} denote the response for the jth subject in the ith cell and let $\boldsymbol{x}_{ij}$ denote the associated $m \times 1$ vector of covariates for this subject. Let $\boldsymbol{Y}$ denote the $n \times 1$ vector of the responses Y_{ij}. Let the matrix $\boldsymbol{W}$ denote the $n \times k$ cell-mean model design matrix and let the matrix $\boldsymbol{X}$ denote the $n \times m$ matrix of covariates. Then symbolically we can write the model as

$$\boldsymbol{Y} = \boldsymbol{W\mu} + \boldsymbol{X\beta} + \boldsymbol{e}, \qquad (5.15)$$

where $\boldsymbol{\mu}$ is the $k \times 1$ vector of means (medians) and $\boldsymbol{\beta}$ is the $m \times 1$ vector of covariate parameters. This is the **adjusted model** and inference for the experimental effects (linear functions of $\boldsymbol{\mu}$) proceeds similar to the inference discussed in Sections 5.2–5.3, except that the inference is adjusted by the covariates. We can also test to see if the covariates make a difference; i.e., test that $\boldsymbol{\beta} = \boldsymbol{0}$. At times, this is of interest.

A major assumption, though, behind the adjusted model (5.15) is that the relationship between the covariates and the response variable is the same at each treatment combination. Thus, before considering adjusted inference, we usually test for no interaction between the experimental design and the covariates. Let $\boldsymbol{Z}$ be the $n \times km$ matrix whose columns are the multiplications, componentwise, of the columns of $\boldsymbol{W}$ and $\boldsymbol{X}$. Then the full model is

$$\boldsymbol{Y} = \boldsymbol{W\mu} + \boldsymbol{X\beta} + \boldsymbol{Z\gamma} + \boldsymbol{e}, \qquad (5.16)$$

where $\boldsymbol{\gamma}$ is the $km \times 1$ vector of interaction parameters. Note that under Model (5.16) each treatment combination has its own linear model with the covariates.

The hypothesis of homogeneous slopes (the linear models of all treatments combinations are the same) is

$$H_{OI} : \boldsymbol{\gamma} = \boldsymbol{0} \text{ versus } H_{AI} : \boldsymbol{\gamma} \neq \boldsymbol{0} \qquad (5.17)$$

Note that studies are often not powered to detect a difference in slopes and a failure to reject the hypothesis (5.17) does not mean the slopes are necessarily the same. If hypotheses (5.17) is rejected, then inference on the effects will often be quite misleading; see, for example, the scatterplot of two groups found in Example 5.4.1. In such cases, confidence intervals for simple contrasts between groups at specified factor values can be carried out. These are often called **pick-a-point** analyses.[5]

5.4.1 Computation of Rank-Based ANCOVA

For the general one-way or k-way ANCOVA models, we have written R functions which compute the rank-based ANCOVA analyses which are included in the R package npsm. We first discuss the computation of rank-based ANCOVA when the design is a one-way layout with k groups.

---

[5]See Huitema (2011) and Watcharotone (2010).

Computation of Rank-Based ANCOVA for a One-Way Layout

For one-way ANCOVA models, we have written two R functions which make use of `Rfit` to compute a rank-based analysis of covariance. The function `onecovaheter` computes the test of homogeneous slopes. It also computes a test of no treatment (group) effect, but this is based on the full model of heterogeneous slopes. It should definitely not be used if the hypothesis of homogeneous slopes is rejected. The second function `onecovahomog` assumes that the slopes are homogeneous and tests the hypothesis of no treatment effect; this is the adjusted analysis of covariance. It also tests the hypotheses that the covariates have a nonsignificant effect. The arguments to these functions are: the number of groups (the number of cells of the one-way design); a $n \times 2$ matrix whose first column contains the response variable and whose second column contains the responses' group identification; and the $n \times m$ matrix of covariates. The functions compute the analysis of covariances, summarizing the tests in ANCOVA tables. These tables are also returned in the value `tab` along with the full model fit in `fit`. For the function `onecovahomog` the full model is (5.15), while for the function `onecovaheter` the full model is (5.16). We illustrate these functions with the following example.

Example 5.4.1 (Chateau Latour Wine Data). Sheather (2009) presents a dataset drawn from the Chateau Latour wine estate. The response variable is the quality of a vintage based on a scale of 1 to 5 over the years 1961 to 2004. The predictor is end of harvest, days between August 31st and the end of harvest for that year, and the factor of interest is whether or not it rained at harvest time. The data are in `latour` in the package `npsm`. We first compute the test for homogeneous slopes.

```
> data = latour[,c('quality','rain')]
> xcov = cbind(latour[,'end.of.harvest'])
> analysis = onecovaheter(2,data,xcov,print.table=T)
```

```
Robust ANCOVA (Assuming Heterogeneous Slopes) Table
                df        RD        MRD        F       p-value
Groups          1 0.8830084 0.8830084 2.395327 0.129574660
Homog Slopes    1 2.8012494 2.8012494 7.598918 0.008755332
```

Based on the robust ANCOVA table, since the p-value is less than 0.01, there is definitely an interaction between the groups and the predictor. Hence, the test in the first row for treatment effect should be ignored.

To investigate this interaction, as shown in the left panel of Figure 5.4, we overlaid the fits of the two linear models over the scatterplot of the data. The dashed line is the fit for the group "rain at harvest time," while the solid line is the fit for the group "no rain at harvest time." For both groups, the quality of the wine decreases as the harvest time increases, but the decrease is much worse if it rains. Because of this interaction the tests in the first two rows of the ANCOVA table are not of much interest. Based on the plot, interpretations from confidence intervals on the difference between the groups at days 25 and 50 would seem to differ.

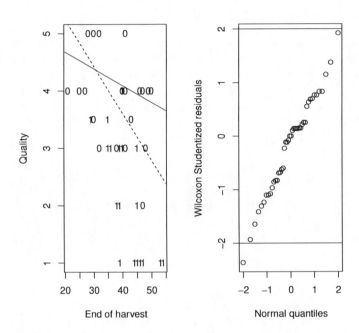

FIGURE 5.4
The left panel is the scatterplot of the data of Example 5.4.1. The dashed line is the fit for the group "rain at harvest time" and the solid line is the fit for the group "no rain at harvest time." The right panel is the normal probability plot of the Wilcoxon Studentized residuals.

∎

If we believe the assumption of equal slopes is a reasonable one, then we may use the function onecovahomog which takes the same arguments as onecovaheter.

Computation of Rank-Based ANCOVA for a k-Way Layout

For the k-way layout, we have written the Rfit function kancova which computes the ANCOVA. Recall that the full model for the design is the cell mean (median) model. Under heterogeneous linear models, each cell in the model has a distinct linear model. This is the full model, Model 5.16, for testing homogeneous slopes. This function also computes the adjusted analysis, assuming that the slopes are homogeneous; that is, for these hypotheses the full model is Model 5.15. So these adjusted tests should be disregarded if there is reason to believe the slopes are different, certainly if the hypothesis of homogeneous slopes is rejected. For the adjusted tests, the standard hypotheses are those of the main effects and interactions of all orders as described in Section 5.3. A test for all covariate effects that are null is also computed. We illustrate this function in the following two examples.

Example 5.4.2. For this illustration we have simulated data from a 2×3 layout with one covariate. Denote the factors by A and B, respectively. It is a null model; i.e., for the data simulated, all effects were set at 0. The data are displayed by cell in the following table. The first column in each cell contains the response realizations while the second column contains the corresponding covariate values.

Factor A	Factor B					
	B(1)		B(2)		B(3)	
A(1)	4.35	4.04	0.69	2.88	4.97	3.4
	5.19	5.19	4.41	5.3	6.63	2.91
	4.31	5.16	7.03	2.96	5.71	3.79
	5.9	1.43	4.14	5.33	4.43	3.53
			4.49	3.51	3.73	4.93
					5.29	5.22
					5.75	3.1
					5.65	3.89
A(2)	4.93	3.22	6.15	3.15	6.02	5.69
	5.1	4.73	4.94	2.01	4.27	4.2
	4.52	2.79	6.1	3.01	4.3	2.57
	5.53	5.63	4.93	3.87	4.47	3.75
	4.21	3.88	5.3	4.47	6.07	2.62
	5.65	3.85				

We use this example to describe the input to the function `kancova`. The design is a two-way with the first factor at 2 levels and the second factor at 3 levels. The first argument to the function is the vector of levels `c(2,3)`. Since there are two factors, the second argument is a matrix of the three columns: vector of responses, Y_{ikj}; level of first factor i; and level of second factor j. The third argument is the matrix of corresponding covariates. The data are in dataset `acov231`. The following code segment computes the rank-based ANCOVA.

```
> levs = c(2,3);
> data = acov231[,1:3];
> xcov = matrix(acov231[,4],ncol=1)
> temp = kancova(levs,data,xcov)
```

Robust ANCOVA Table
All tests except last row is with homogeneous slopes
as the full model.   For the last row the full model is
with heteroscedastic slopes.

	df	RD	MRD	F	p-value
1 , 0	1	0.21806168	0.21806168	0.46290087	0.5022855
0 , 1	2	0.09420198	0.04710099	0.09998588	0.9051964
1 , 1	2	1.21200859	0.60600430	1.28642461	0.2932631
Covariate	1	0.13639223	0.13639223	0.28953313	0.5950966
Hetrog regr	5	2.33848642	0.46769728	0.98946073	0.4476655

The rank-based tests are all nonsignificant, which agrees with the null model used to generate the data. ■

Example 5.4.3 (2 × 2 with Covariate). Huitema (2011), page 496, presents an example of a 2 × 2 layout with a covariate. The dependent variable is the number of novel responses under controlled conditions. The factors are type of reinforcement (Factor A at 2 levels) and type of program (Factor B at 2 levels); hence there are four cells. The covariate is a measure of verbal fluency. There are only 4 observations per cell for a total of $n = 16$ observations. Since there are 8 parameters in the heterogeneous slope model, there are only 2 observations per parameter. Hence, the results are tentative. The data are in the dataset huitema496. Using the function kancova with the default Wilcoxon scores, the following robust ANCOVA table is computed.

```
> levels = c(2,2);
> y.group = huitema496[,c('y','i','j')]
> xcov = huitema496[,'x']
> temp = kancova(levels,y.group,xcov)
```

Robust ANCOVA Table
All tests except last row is with homogeneous slopes
as the full model. For the last row the full model is
with heteroscedastic slopes.

	df	RD	MRD	F	p-value
1 , 0	1	5.6740175	5.6740175	6.0883935	0.031261699
0 , 1	1	0.4937964	0.4937964	0.5298585	0.481873895
1 , 1	1	0.1062181	0.1062181	0.1139752	0.742017556
Covariate	1	12.2708792	12.2708792	13.1670267	0.003966071
Hetrog regr	3	8.4868988	2.8289663	4.2484881	0.045209629

The robust ANCOVA table indicates heterogeneous slopes, so we plotted the four regression models next as shown in Figure 5.5. The rank-based test of homogeneity of the sample slopes agrees with the plot. In particular, the slope for the cell with $A = 2, B = 2$ differs from the others. Again, these results are based on small numbers and thus should be interpreted with caution. As a pilot study, these results may serve in the conduction of a power analysis for a larger study. ■

The rank-based tests computed by these functions are based on reductions of dispersion as we move from reduced to full models. Hence, as an alternative

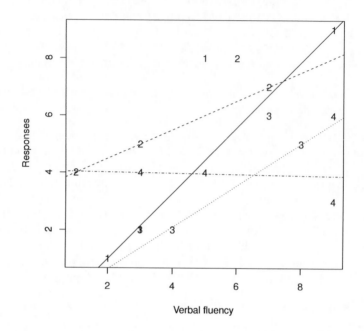

FIGURE 5.5
Scatterplot of the data of Example 5.4.3. The number 1 represents observations from the cell with $A = 1, B = 1$, 2 represents observations from the cell with $A = 1, B = 2$, 3 represents observations from the cell with $A = 2, B = 1$, and 4 represents observations from the cell with $A = 2, B = 2$.

to these functions, the computations can also be obtained using the Rfit functions rfit and drop.test with only a minor amount of code. In this way, specific hypotheses of interest can easily be computed, as we show in the following example.

Example 5.4.4 (Triglyceride and Blood Plasma Levels). The data for this example are drawn from a clinical study discussed in Hollander and Wolfe (1999). The data consist of triglyceride levels on 13 patients. Two factors, each at two levels, were recorded: Sex and Obesity. The concomitant variables are chylomicrons, age, and three lipid variables (very low-density lipoproteins (VLDL), low-density lipoproteins (LDL), and high-density lipoproteins (HDL)). The data are in the npsm dataset blood.plasma. The next code segment displays a subset of it.

```
> head(blood.plasma)
```
```
     Total Sex Obese Chylo VLDL  LDL  HDL Age
```

```
[1,] 20.19   1     1  3.11 4.51 2.05 0.67  53
[2,] 27.00   0     1  4.90 6.03 0.67 0.65  51
[3,] 51.75   0     0  5.72 7.98 0.96 0.60  54
[4,] 51.36   0     1  7.82 9.58 1.06 0.42  56
[5,] 28.98   1     1  2.62 7.54 1.42 0.36  66
[6,] 21.70   0     1  1.48 3.96 1.09 0.23  37
```

The design matrix for the full model to test the hypotheses of no interaction between the factors and the covariates would have 24 columns, which, with only 13 observations, is impossible. Instead, we discuss the code to compute several tests of hypotheses of interest. The full model design matrix consists of the four dummy columns for the cell means and the 5 covariates. In the following code segment this design matrix is in the R matrix xfull while the response, total triglyceride, is in the column Total of blood.plasma. The resulting full model fit and its summary are given by:

```
> fitfull = rfit(blood.plasma[,'Total']~xfull-1)
> summary(fitfull)

Call:
rfit.default(formula = blood.plasma[, "Total"] ~ xfull - 1)

Coefficients:
            Estimate Std. Error t.value  p.value
xfull00      8.59033    9.30031  0.9237 0.407938
xfull01     -3.00427    8.32005 -0.3611 0.736297
xfull10    -12.61631   10.11257 -1.2476 0.280234
xfull11    -11.58851   10.32710 -1.1221 0.324605
xfullChylo   1.74111    0.53220  3.2715 0.030745 *
xfullVLDL    2.87822    0.41674  6.9064 0.002305 **
xfullLDL     3.79748    2.77105  1.3704 0.242433
xfullHDL   -11.46968    4.61116 -2.4874 0.067674 .
xfullAge     0.24942    0.13359  1.8671 0.135284
---
Signif. codes:  0 '***' 0.001 '**' 0.01 '*' 0.05 '.' 0.1 ' ' 1

Multiple R-squared (Robust): 0.9655414
Reduction in Dispersion Test: 14.01015 p-value: 0.01108

> dfull = disp(fitfull$betahat, xfull, fitfull$y, fitfull$scores)
> dfull

          [,1]
[1,] 26.44446
```

The last line is the minimum value of the dispersion function based on the fit of the full model. The one hypothesis of interest discussed in Hollander and

Wolfe is whether or not the three lipid covariates (VLDL, LDL, and HDL) are significant predictors. The answer appears to be "yes" based on the above summary of the Rfit of the full model. The following code performs a formal test using the hypothesis matrix hmat:

```
> hmat = rbind(c(rep(0,5),1,rep(0,3)),
+             c(rep(0,6),1,rep(0,2)),
+             c(rep(0,7),1,rep(0,1)))
> hmat

      [,1] [,2] [,3] [,4] [,5] [,6] [,7] [,8] [,9]
[1,]    0    0    0    0    0    1    0    0    0
[2,]    0    0    0    0    0    0    1    0    0
[3,]    0    0    0    0    0    0    0    1    0

> xred1 = redmod(xfull,hmat)
> fitr1 = rfit(blood.plasma[,'Total']~xred1-1)
> drop.test(fitfull,fitr1)

Drop in Dispersion Test
F-Statistic      p-value
  10.407321      0.023246
```

Hence, based on this *p*-value (0.0232), it seems that the lipid variables are related to triglyceride levels. The next few lines of code test to see if the factor sex has an effect on triglycerides.

```
> hmat=rbind(c(1,1,-1,-1,rep(0,5)))
> xred3 = redmod(xfull,hmat)
> fitr3 = rfit(blood.plasma[,'Total']~xred3-1)
> drop.test(fitfull,fitr3)

Drop in Dispersion Test
F-Statistic      p-value
  21.79352       0.00953
```

Based on the *p*-value of 0.0095, it appears that the factor sex also has an effect on triglyceride levels. Finally, consider the effect of obesity on triglyceride level.

```
> hmat=rbind(c(1,-1,1,-1,rep(0,5)))
> xred2 = redmod(xfull,hmat)
> fitr2 = rfit(Total~xred2-1)
> drop.test(fitfull,fitr2)

Drop in Dispersion Test
F-Statistic      p-value
  11.580682      0.027201
```

Thus, the rank-based test for obesity results in the test statistic 11.58 with *p*-value 0.0272. ∎

5.5   Methodology for Type III Hypotheses Testing[*]

In this section, we briefly describe how `Rfit` obtains Type III hypotheses for the two-way and k-way designs. Consider first the hypotheses for the two-way design which are given in expressions (5.11)–(5.13). For our discussion, assume that the data are stacked as the $n \times 1$ vector $\boldsymbol{Y}$ by cell and row-by-row in row order; i.e., in terms of the subscripts ijk, k runs the fastest and i runs the slowest. Let $\boldsymbol{\mu}$ denote the corresponding $ab \times 1$ vector of parameters and let $\boldsymbol{W}$ denote the $n \times ab$ incidence matrix. Then the full model can be written as $\boldsymbol{Y} = \boldsymbol{W}\boldsymbol{\mu} + \boldsymbol{e}$, where $\boldsymbol{e}$ denotes the vector of random errors.

For the two-way model, the three hypotheses of interest are the main effects hypotheses and the interaction hypothesis given respectively by (5.11)–(5.13). Following Hocking (1985) the hypotheses matrices $\boldsymbol{M}$ can easily be computed in terms of Kronecker products. For a positive integer s, define the augmented matrix

$$\boldsymbol{\Delta}_s = [\boldsymbol{I}_{s-1} \; -\boldsymbol{1}_{s-1}], \tag{5.18}$$

where $\boldsymbol{I}_{s-1}$ is the identity matrix of order $s-1$ and $\boldsymbol{1}_{s-1}$ denotes a vector of $(s-1)$ ones. For our two-way design with A at a levels and B at b levels, the hypothesis matrices of average main effects and interaction are given by

$$\text{For Hypothesis (5.11): } \boldsymbol{M}_A \;=\; \boldsymbol{\Delta}_a \otimes \frac{1}{b}\boldsymbol{1}_b^T$$

$$\text{For Hypothesis (5.12): } \boldsymbol{M}_B \;=\; \frac{1}{a}\boldsymbol{1}_a^T \otimes \boldsymbol{\Delta}_b$$

$$\text{For Hypothesis (5.13): } \boldsymbol{M}_{A\times B} \;=\; \boldsymbol{\Delta}_a \otimes \boldsymbol{\Delta}_b,$$

where $\otimes$ denotes the Kronecker product. Based on these hypothesis matrices, `Rfit` computes[6] reduced model design matrices.

Hypothesis matrices for higher order designs can be computed[7] similarly. For example, suppose we have the four factors A, B, C and D with respective levels a, b, c and d. Then the hypotheses matrices to test the interaction between B and D and the 4-way interaction are respectively given by

$$\boldsymbol{M}_{B\times D} \;=\; \frac{1}{a}\boldsymbol{1}_a^T \otimes \boldsymbol{\Delta}_b \otimes \frac{1}{c}\boldsymbol{1}c^T \otimes \boldsymbol{\Delta}_d$$

$$\boldsymbol{M}_{A\times B\times C'B\times D} \;=\; \boldsymbol{\Delta}_a \otimes \boldsymbol{\Delta}_b \otimes \boldsymbol{\Delta}_c \otimes \boldsymbol{\Delta}_d.$$

The corresponding reduced model design matrices can be easily computed to obtain the tests of the hypotheses.

---

[6]See page 209 of Hettmansperger and McKean (2011).
[7]See Hocking (1985).

5.6   Ordered Alternatives

Consider again the one-way ANOVA model, Section 5.2 with k levels of a factor or k treatments. The full model is given in expression (5.1). As in Section 5.2, let Y_{ij} denote the jth response in sample i for $j = 1, \ldots, n_i$ and $i = 1, \ldots, k$, and let μ_i denote the mean or median of Y_{ij}. In this section, suppose that a certain ordering of the means (centers) is a reasonable alternative to consider; that is, assume that the hypotheses are

$$H_0 : \mu_1 = \cdots = \mu_k \text{ vs. } H_A : \mu_1 \leq \mu_2 \leq \cdots \leq \mu_k, \qquad (5.19)$$

with at least one strict inequality in H_A. There are several procedures designed to test these hypotheses. One of the most popular procedures is the distribution-free Jonckheere–Terpstra test. The test statistic is the sum of pairwise Mann–Whitney statistics over samples $u < v$. That is, let

$$U_{uv} = \#_{1 \leq j \leq n_u, 1 \leq l \leq n_v}\{Y_{ju} < Y_{lv}\}, \quad 1 \leq u < v \leq k. \qquad (5.20)$$

Then the Jonckheere–Terpstra test statistic is defined by

$$J = \sum_{u=1}^{k-1} \sum_{v=2}^{k} U_{uv}. \qquad (5.21)$$

The null hypothesis H_0 is rejected in favor of H_A, ordered alternatives, for large values of J.

The statistic J is distribution-free and there are some tables[8] for its critical values in the literature. Often the asymptotic standardized test statistic is used. The mean and variance under the null hypothesis as well as the standardized test statistic of J are given by

$$
\begin{aligned}
E_{H_0}(J) &= \frac{n^2 - \sum_{i=1}^{k} n_i^2}{4} \\[1em]
V_{H_0}(J) &= \frac{n^2(2n+3) - \sum_{i=1}^{k} n_i^2(2n_i+3)}{72} \\[1em]
z_J &= \frac{J - E_{H_0}(J)}{\sqrt{V_{H_0}(J)}}.
\end{aligned}
\qquad (5.22)
$$

An asymptotic level $0 < \alpha < 1$ test is to reject H_0 if $z_J \geq z_\alpha$.

In the case of tied observations, the contribution to U_{uv}, (5.20), is one-half. That is, if $Y_{ju} = Y_{lv}$ then instead of the contribution of 0 to U_{uv} as expression (5.20) dictates for this pair, the contribution is one-half. If there are many ties then the tie correction to the variance[9] should be used. Included in npsm

---

[8]See Section 6.2 of Hollander and Wolfe (1999).
[9]See page 204 of Hollander and Wolfe (1999).

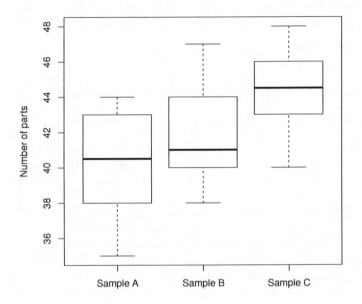

FIGURE 5.6
Comparison boxplots of three samples of Example 5.6.1.

is the function `jonckheere`. This software adjusts the null variance for ties, returning the test statistic and its asymptotic p-value. We illustrate its use in the following example.

Example 5.6.1 (Knowledge of Performance). Hollander and Wolfe (1999) discuss a study on workers whose job was a repetitive task. It was thought that knowledge of performance would improve their output. So 18 workers were randomly divided into three groups A, B and C. Group A served as a control. These workers were given no information about their performance. The workers in Group B received some information about their performance, while those in Group C were given detailed information on their performance. The response was the number of parts each worker produced in the specified amount of time. The data appear in the next table. Figure 5.6 shows comparison boxplots of the samples.

	Parts Produced					
Group A	40	35	38	43	44	41
Group B	38	40	47	44	40	42
Group C	48	40	45	43	46	44

The alternative hypothesis is $H_A : \mu_A \leq \mu_b \leq \mu_c$ with at least one strict inequality, where μ_i is the mean output for a worker in Group i. In the next code segment, the arguments in the call jonckheere(response,indicator) are the vector of responses and the vector of group indicators.

```
> jonckheere(response,indicator)
```

```
    Jonckheere         ExpJ          VarJ              P
  79.00000000   54.00000000  150.28676471   0.02071039
```

With the p-value$=0.0207$ there is evidence to support the alternative hypothesis, confirming the boxplots. ∎

As the reader is asked to show in Exercise 5.8.3, for the data of the last example, the Kruskal–Wallis test statistic has the value $H = 4.362$ with p-value 0.1130, which is not significant at the 5% level. The Kruskal–Wallis test is generally less powerful than the Jonckheere–Terpstra test for ordered alternatives; for verification see the small study requested in Exercise 5.8.10.

There are other distribution-free tests for ordered alternatives. One such procedure is based on Spearman's correlation coefficient r_s defined in expression (4.37). For its application, using the above notation, let $X_{ij} = i$, i.e., the group indicator. Then the test[10] is based on Spearman's correlation coefficient between X_{ij} and Y_{ij}. This test is distribution-free. For inference, recall that under the null distribution, $z_s = \sqrt{n}r_s$ is asymptotically $N(0,1)$. Hence, z_s can be used as a standardized test statistic. A degree of freedom correction[11], though, makes use of the standardized test statistic t_s which is defined by

$$t_s = \frac{r_s}{\sqrt{(1 - r_s^2)/(n-2)}}. \tag{5.23}$$

We use the test statistics t_s in the next code segment to compute this Spearman procedure for the data in the last example.

```
> cor.test(indicator,response,method="spearman",
+       continuity=FALSE,exact=FALSE,alternative='greater')
```

```
          Spearman's rank correlation rho

data:  indicator and response
S = 488.4562, p-value = 0.01817
alternative hypothesis: true rho is greater than 0
sample estimates:
      rho
0.4959172
```

---

[10] See Tryon and Hettmansperger (1973) and McKean et al. (2001) for details.
[11] See page 347 of Huitema (2011).

Thus the tests based on Spearman's ρ_S and the Jonckheere–Terpstra are essentially in agreement for this example.

McKean et al. (2001) developed a bootstrap procedure for the Spearman procedure. Note that r_s is an estimate of association and the confidence interval is a measure of the strength of this association.

5.7   Multi-Sample Scale Problem

A general assumption in fixed-effects ANOVA-ANCOVA is the homogeneity of scale of the random errors from level (group) to level (group). For two levels (samples), we discussed the Fligner–Killeen test for homogeneity of scale in Section 3.3. This test generalizes immediately to the k-level (sample) problem.

For discussion, we use the notation of the k-cell (sample) model in Section 5.2 for the one-way ANOVA design. For this section, our full model is Model (5.1) except that the pdf of random errors for the jth level is of the form $f_j(x) = f[(x - \theta_j)/\sigma_j]/\sigma_j$, where θ_j is the median and $\sigma_j > 0$ is a scale parameter for the jth level, $j = 1, \ldots, k$. A general hypothesis of interest is that the scale parameters are the same for each level, i.e.,

$$H_0 : \sigma_1 = \cdots = \sigma_k \text{ versus } H_A : \sigma_j \neq \sigma_{j'} \text{ for some } j \neq j'. \qquad (5.24)$$

Either this could be the hypothesis of interest or the hypothesis for a pre-test on homogeneous scales in the one-way ANOVA location model.

The Fligner–Killeen test of scale for two-samples, (3.30), easily generalizes to this situation. As in Section 3.3, define the folded-aligned sample as

$$Y_{ij}^* = Y_{ij} - \text{med}_{j'}\{Y_{ij'}\}, \quad j = 1, \ldots, n_i : i = 1, \ldots, k. \qquad (5.25)$$

Let $n = \sum_{i=1}^{n} n_i$ denote the total sample size and let $R_{ij} = R|Y_{ij}^*|$ denote the rank of the absolute values of these items, from 1 to n. Define the scores $a^*(i)$ as

$$a(l) = \left[\Phi^{-1}\left(\frac{l}{2(n+1)} + \frac{1}{2} \right) \right]^2$$

$$\bar{a} = \frac{1}{n} \sum_{l=1}^{n} a(l)$$

$$a^*(l) = a(l) - \bar{a}. \qquad (5.26)$$

Then the Fligner–Killeen test statistic is

$$Q_{FK} = \frac{n-1}{\sum_{l=1}^{n}(a^*(l))^2} \sum_{i=1}^{k} \left\{ \sum_{j=1}^{n_i} a^*(R_{ij}) \right\}^2. \qquad (5.27)$$

An approximate level α-test[12] is based on rejecting H_0 if $Q_{FK} \geq \chi_\alpha^2(k-1)$.

As in Section 3.3, we can obtain rank-based estimates of the difference in scale. Let $Z_{ij} = \log(|Y_{i1}^*|)$. and let $\Delta_{1i} = \log(\sigma_i/\sigma_1)$, for $i = 2, \ldots, k$, where, without loss of generality, we have referenced the first level. Using $\Delta_{11} = 0$, we can write the log-linear model for the aligned, folded sample as

$$Z_{ij} = \Delta_{1i}^* + e_{ij}, \quad j = 1, \ldots n_i, i = 1, 2, \ldots, k. \tag{5.28}$$

As discussed in Section 3.3, the scores defined in expression (3.29) are appropriate for the rank-based fit of this model. Recall that they are optimal, when the random errors of the original samples are normally distributed. As discussed in Section 3.3, exponentiation of the regression estimates leads to estimation (and confidence intervals) for the ratio of scales $\eta_{1i} = \sigma_i/\sigma_1$.

The function fkk.test is a wrapper which obtains the R fit and analysis for this section. We demonstrate it in the following example.

Example 5.7.1 (Three Generated Samples). For this example, we generated three samples (rounded) from Laplace distributions. The samples have location and scale $(5, 1)$, $(10, 4)$, and $(10, 8)$ respectively. Hence in the above notation, $\eta_{21} = 4$ and $\eta_{31} = 8$. A comparison boxplot of the three samples is shown in Figure 5.7.

The following code segment computes the Fligner–Killeen test for these three samples. Note the response variables are in the vector response and the vector indicator is a vector of group membership.

```
> fkk.test(response,indicator)
```

```
Table of estimates and  95  percent confidence intervals:
                    estimate  ci.lower ci.upper
xmatas.factor(iu)2 3.595758 0.9836632 13.14421
xmatas.factor(iu)3 8.445785 2.5236985 28.26458

 Test statistic =  8.518047  p-value =  0.0141361
```

Hence, based on the results of the test, there is evidence to reject H_0. The estimates of η_{21} and η_{31} are close to their true values. The respective confidence intervals are $(0.98, 13.14)$ and $(2.52, 28.26)$. ∎

5.8 Exercises

5.8.1. Hollander and Wolfe (1999) report on a study of the length of YOY gizzard shad fish at four different sites of Kokosing Lake in the summer of 1984. The data are:

---

[12]See page 105 of Hájek and Šidák (1967).

Site 1	Site 2	Site 3	Site 4
46	42	38	31
28	60	33	30
46	32	26	27
37	42	25	29
32	45	28	30
41	58	28	25
42	27	26	25
45	51	27	24
38	42	27	27
44	52	27	30

Let μ_i be the true mean length of YOY gizzard shad at site i.

(a) Use the rank-based Wilcoxon procedure, (5.7), to test the hypothesis of equal means.

(b) Based on Part (a), use Fisher's least significance difference to perform a multiple comparison on the differences in the means. As discussed in Hollander and Wolfe, YOY gizzard shad are eaten by

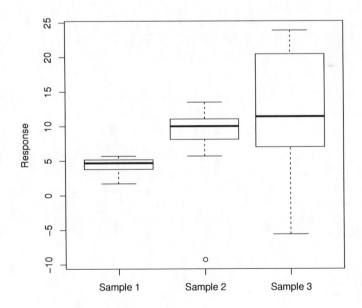

FIGURE 5.7
Comparison boxplots of three samples.

game fish and for this purpose smaller fish are better. In this regard, based on the MCP analysis, which sites, if any, are preferred?

5.8.2. For the study discussed in Example 5.2.3, obtain the analysis based on the Wilcoxon test using F_W, (5.7). Then obtain the MCP analysis using Tukey's method. Compare this analysis with the Kruskal–Wallis analysis presented in the example.

5.8.3. For the data of Example 5.6.1 determine the value of the Kruskal–Wallis test and its p-value.

5.8.4. For a one-way design, instead of using the `oneway.rfit` function, the rank-based test based on F_W, expression (5.7), can be computed by the `Rfit` functions `rfit` and `drop.test`.

(a) Suppose the R vector `y` contains the combined samples and the R vector `ind` contains the associated levels. Discuss why following two code segments are equivalent:
`oneway.rfit(y,ind)` and
`fit<-rfit(y~factor(ind)); drop.test(fit)`

(b) Verify this equivalence for the data in Example 5.2.1.

5.8.5. Write an R script which compares empirically the power between the rank-based test based on F_W and the corresponding LS test for the following situation: 4 samples of size 10; location centers 10, 11, 12, and 15; and the random errors $3 * e_{ij}$ where the e_{ij} are iid with the common t-distribution having 3 degrees of freedom. Use a simulation size of 1000 and the level $\alpha = 0.05$.

5.8.6. In Exercise 5.8.5, we simulated the empirical power of the rank-based and LS tests for a specific situation. For this exercise, check the validity of the rank-based and LS tests; i.e., set the location centers to be the same.

5.8.7. Suppose that we want a descriptive plot for a one-way design. Comparison boxplots are one such plot; however, if the level sample sizes n_i are small then these plots can be misleading (quartiles and, hence, lengths of boxplots, can be adversely affected by a few outliers). Hence, for $n_i \leq 10$, we recommend comparison dotplots instead of boxplots. Consider the following data from a one-way design.

Level 1	66	45	42	53	71		
Level 2	38	53	47	23	42	50	
Level 3	82	26	95	70	80	82	75

(a) Obtain the comparison dotplots for the above data.

(b) Compute the Fligner–Killeen test of equal scales for these data.

5.8.8. Miliken and Johnson (1984) discuss a study pertaining to an unbalanced 2×3 crossed factorial design. For convenience, we present the data below. For their LS analysis, Milliken and Johnson recommend Type III hypotheses. As discussed in Section 5.5, the rank-based analysis based on the Rfit function raov obtains tests based on Type III hypotheses. The R function lm, however, does not. In this exercise, we show how to easily obtain LS analyses for Type III hypotheses using the functions redmod and cellx from Rfit . For the data, the factors are labeled T and B and the responses are tabled as:

	B_1	B_2	B_3
T_1	19	24	22
	20	26	25
	21		25
T_2	25	21	31
	27	24	32
		24	33

The code below assumes that the response and indicator vectors are:

```
resp = c(19, 20, 21, 24, 26, 22, 25, 25, 25, 27, 21, 24, 24,
         31, 32, 33)
a = c(rep(1,8),rep(2,8))
b = c(1, 1, 1, 2, 2, 3, 3, 3, 1, 1, 2, 2, 2, 3, 3, 3)
```

(a) First obtain the analysis for interaction and main effects using the Rfit function raov. The hypotheses of this analysis are of Type III.

(b) The following script will obtain the LS Type III analysis for Factor T:

```
fitls <- lm(resp ~factor(a):factor(b))
cell <- rep(1:6,each=3)
cellmean <- cellx(cell)
ha <- c(1,1,1,-1,-1,-1)
xa <- redmod(cellmean,ha)
lmred <- lm(resp ~ xa)
anova(lmred,fitls)
```

Run this code and show that the LS test statistic computes to $F = 30.857$. Notice that this differs from the LS ANOVA based on fitls.

(c) Write code and run it for the LS Type III analysis of Factor B.

Hint: the hypothesis matrix hb has two rows.

(d) Write code and run it for the LS Type III analysis of interaction.

Hint: the hypothesis matrix hint has two rows. Notice that it agrees with the LS ANOVA based on fitls.

5.8.9. Page 436 of Hollander and Wolfe (1999), presents part of a study on the effects of cloud seeding on cyclones; see Wells and Wells (1967) for the original reference. For the reader's convenience, the data are contained in the dataset SCUD. The first column is an indicator for Control (2) or Seeded (1); column 2 is the predictor M, the geostrophic meridional circulation index; and column 3 is the response RI which is a measure of precipitation.

    (a)   Obtain a scatterplot of RI versus M. Use different plotting symbols for the Control and Seeded. Add the rank-based fits of the linear models for each. Comment on the plot.

    (b)   Using a rank-based analysis, test for homogeneous slopes for the two groups.

    (c)   If homogeneous slopes is "accepted" in (b), use a rank-based analysis to test for homogeneous groups.

    (d)   The test in Part (c) is adjusted for M. Is this adjustment necessary? Test at level $\alpha = 0.05$.

5.8.10. Using a simulation study investigate the powers of the Jonckheere–Terpstra test and the Kruskal–Wallis test for the following situation: Samples of size 10 from four normal populations each having variance 1 and with the respective means of $\mu_1 = 0$, $\mu_2 = 0.45$, $\mu_3 = .90$, and $\mu_4 = 1.0$. Use the level of $\alpha = 0.05$ and a simulation size of 10,000.

5.8.11. In reading through Section 5.6 on ordered alternatives, the reader may have noticed the simplicity of the test based on Spearman's ρ_S over the test using the Jonckheere–Terpstra test statistic. Is it as powerful? As a partial answer, this exercise provides some empirical evidence. One may use cor.test to obtain a test based on Spearman's ρ. See, for example, the following code.

```
group <- c(rep(1,ni),rep(2,ni),rep(3,ni),rep(4,ni))
y1 <- rnorm(ni,0,1);y2 <- rnorm(ni,.15,1);
y3 <- rnorm(ni,.35,1); y4 <- rnorm(ni,.55,1)
y <- c(y1,y2,y3,y4)
cor.test(group,y,method='spearman',
    continuity=FALSE,exact=FALSE,alternative='less')
```

    (a)   Determine the situation (distributions, alternative, etc.) which the above code simulates.

    (b)   Based on the above situation, run a simulation to compare the empirical powers of the Jonckheere–Terpstra test and Spearman's ρ.

    (c)   Run a simulation where the error distribution is a t-distribution with 3 degrees of freedom.

    (d)   Run a simulation where the error distribution is a χ^2-distribution with 1.5 degrees of freedom.

5.8.12. Besides simplicity, another advantage of the test based Spearman's ρ_S over the Jonckheere–Terpstra is that the estimate of ρ_S is an easily understood correlational measure. In this setting, it offers a measure of the "strength" of the relationship. Use the function `cor.boot.ci` to obtain a bootstrap confidence interval.

5.8.13. Consider the malignant melanoma data in Example 3.1.1. See if the association found there still holds after adjusting for latitude and longitude.

6

Time to Event Analysis

6.1 Introduction

In survival or reliability analysis the investigator is interested in time to an event of interest as the outcome variable. Often in a clinical trial the goal is to evaluate the effectiveness of a new treatment at prolonging survival; i.e. to extend the time to the event of death. It is usually the case that at the end of followup a portion of the subjects in the trial have not experienced the event; for these subjects the outcome variable is **censored**. Similarly, in engineering studies, often the lifetimes of mechanical or electrical parts are of interest. In a typical experimental design, lifetimes of these parts are recorded along with covariates (including design variables). Often the lifetimes are called failure times, i.e., times until failure. As in a clinical study, at the end of the experiment, there may be parts which are still functioning (censored observations).

In this chapter we discuss standard nonparametric and semiparametric methods for analysis of time to event data. In Section 6.2, we discuss the Kaplan–Meier estimate of the survival function for these models and associated nonparametric tests. Section 6.3 introduces the proportional hazards analysis for these models, while in Section 6.4 we discuss rank-based fits of accelerated failure time models, which include proportional hazards models. We illustrate our discussion with analyses of real datasets based on computation by R functions. For a more complete introduction to survival data we refer the reader to Chapter 7 of Cook and DeMets (2008) or to the monograph by Kalbfleisch and Prentice (2002). Therneau and Grambsch (2000) provide a thorough treatment of modeling survival data using SAS and R/S.

6.2 Kaplan–Meier and Log Rank Test

Let T denote the time to an event. Assume T is a continuous random variable with cdf $F(t)$. The survival function is defined as the probability that a subject survives until at least time t; i.e., $S(t) = P(T > t) = 1 - F(t)$. When all subjects in the trial experience the event during the course of the study,

TABLE 6.1

Survival Times (in months) for Treatment of
Pulmonary Metastasis.

11	13	13	13	13	13	14	14	15	15	17

so that there are no censored observations, an estimate of $S(t)$ may be based on the empirical cdf. However, in most studies there are a number of subjects who are not known to have experienced the outcome prior to the study completion. Kaplan and Meier (1958) developed their product-limit estimate as an estimate of $S(t)$ which incorporates information from censored observations. In this section we briefly discuss estimates of the survival function and also illustrate them via small samples. The focus, however, is on the R syntax for analysis. We describe how to store time to event data and censoring in R, as well as computation of the Kaplan–Meier estimate and the log-rank test – which is a standard test for comparing two survival distributions.

We begin with a brief overview of survival data as well as simple examples which illustrate the calculation of the Kaplan–Meier estimate.

Example 6.2.1 (Treatment of Pulmonary Metastasis). In a study of the treatment of pulmonary metastasis arising from osteosarcoma, survival time was collected; the data are provided in Table 6.1.

As there are no censored observation an estimate of the survival function at time t is

$$\hat{S}(t) = \frac{\#\{t_i > t\}}{n} \tag{6.1}$$

which is based on the empirical cdf. Because of the low number of distinct time points the estimate (6.1) is easily calculated by hand which we briefly illustrate next. Since $n = 11$, the result is

$$\hat{S}(t) = \begin{cases} 1 & 0 \leq t < 11 \\ \frac{10}{11} & 11 \leq t < 13 \\ \frac{5}{11} & 13 \leq t < 14 \\ \frac{3}{11} & 14 \leq t < 15 \\ \frac{1}{11} & 15 \leq t < 17 \\ 0 & t \geq 17. \end{cases}$$

The estimated survival function is plotted in Figure 6.1. ∎

Though (6.1) aids in the understanding of survival functions, it is not often useful in practice. In most clinical studies, at the end of followup there are subjects who have yet to experience the event being studied. In this case, the Kaplan–Meier product limit estimate is used which we describe briefly next. Suppose n experimental units are put on test. Let $t_{(1)} < \ldots < t_{(k)}$ denote the ordered distinct event times. If there are censored responses, then $k < n$. Let

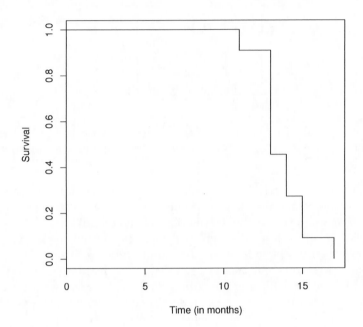

FIGURE 6.1
Estimated survival curve ($\hat{S}(t)$).

n_i = #subjects at risk at the beginning of time $t_{(i)}$ and d_i = #events occurring at time $t_{(i)}$ (i.e., during that day, month, etc.). The **Kaplan–Meier** estimate of the survival function is defined as

$$\hat{S}(t) = \prod_{t_{(i)} \leq t} \left(1 - \frac{d_i}{n_i}\right). \tag{6.2}$$

Note that when there is no censoring (6.2) reduces to (6.1). To aid in interpretation, we illustrate the calculation in the following example.

Example 6.2.2 (Cancer Remission: Time to Relapse.). The data in Table 6.2 represent time to relapse (in months) in a cancer study. Notice, based on the

TABLE 6.2
Time in Remission (in months) in Cancer Study.

Relapse		3	6.5	6.5	10	12	15
Lost to followup	8.4						
Alive and in remission at at end of study		4	5.7	10			

TABLE 6.3

Illustration of the Kaplan–Meier Estimate.

t	n	d	$1 - d/n$	$S(t)$
3	10	1	$9/10 = 0.9$	0.9
6.5	7	2	$5/7 = 0.71$	$0.9*0.71 = 0.64$
10	4	1	$3/4 = 0.75$	$0.64*0.75 = 0.48$
12	2	1	$1/2 = 0.5$	$0.48*0.5 = 0.24$
15	1	1	$0/1 = 0.0$	0

top row of the table, that there are $k = 5$ distinct survival event times. Table 6.3 illustrates the calculation of of the Kaplan–Meier estimate for this dataset. ∎

Often a study on survival involves the effect that different treatments have on survival time. Suppose we have r independent groups (treatments). Let H_0 be the null hypothesis that the distributions of the groups are the same; i.e., the population survival functions are the same. Obviously, overlaid Kaplan–Meier survival curves provide an effective graphical comparison of the times until failure of the different treatment groups. A nonparametric test that is often used to test for a difference in group survival times is the log-rank test. This test is complicated and complete details can be found, for example, in Kalbfleisch and Prentice (2002). Briefly, as above, let $t_1 < t_2 < \cdots < t_k$ be the distinct failure times of the combined samples. Then at each time point t_j, it can be shown that the number of failures in Group i conditioned on the total number of failures has a distribution-free hypergeometric distribution under H_0. Based on this a goodness-of-fit type test statistic (called the log-rank test) can be formulated which has a χ^2-distribution with $r - 1$ degrees of freedom under H_0. The next example illustrates this discussion for the time until relapse of two groups of patients who had survived a lobar intracerebral hemorrhage.

Example 6.2.3 (Hemorrhage Data). For demonstration we use the hemorrhage data discussed in Chapter 6 of Dupont (2002). The study population consisted of patients who had survived a lobar intracerebral hemorrhage and whose genotype was known. The outcome variable was the time until recurrence of lobar intracerebral hemorrhage. The investigators were interested in examining the genetic effect on recurrence as there were three common alleles $e2, e3$, and $e4$. The analysis was focused on the effect of homozygous $e3/e3$ (Group 1) versus at least one $e2$ or $e4$ (Group 2). The data are available at the author's website. The following code segment illustrates reading the data into R and converting it to a survival dataset which includes censoring information. Many of the functions for survival data are available in the R package survival (Therneau 2013).

```
> with(hemorrhage,Surv(round(time,2),recur))
```

```
 [1]  0.23    1.05+  1.22    1.38+  1.41    1.51+  1.58+  1.58    3.06   3.32
[11]  3.52    3.55   4.04+   4.63+  4.76    8.08+  8.44+  9.53   10.61+ 10.68+
[21] 11.86+  12.32  13.27+  13.60+ 14.69+  15.57  16.72+ 17.84+ 18.04+ 18.46+
[31] 18.46+  18.46+ 18.66+  19.15  19.55+  19.75+ 20.11+ 20.27+ 20.47+ 24.77
[41] 24.87   25.56+ 25.63+  26.32+ 26.81+  28.09  30.52+ 32.95+ 33.05+ 33.61
[51] 34.99+  35.06+ 36.24+  37.03+ 37.52   37.75+ 38.54+ 38.97+ 39.16+ 40.61+
[61] 42.22+  42.41+ 42.78+  42.87  43.27+  44.65+ 45.24+ 46.29+ 46.88+ 47.57+
[71] 53.88+
```

In the output are survival times (in months) for 71 subjects. However, one subject's genotype information is missing and is excluded from analysis. Of the remaining 70 subjects, 32 are in Group 1 and 38 are in Group 2. A + sign indicates a censored observation; meaning that at that point in time the subject had yet to report recurrence. The study could have ended or the subject could have been lost to followup. Kaplan–Meier estimates are available through the command `survfit`. The resulting estimates may then be plotted, as is usually the case for Kaplan–Meier estimates, as the following code illustrates. If confidence bands are desired, one may use the `conf.type` option to survfit. Setting `conf.type='plain'` returns the usual Greenwood (1926) estimates.

```
> fit<-with(hemorrhage, survfit(Surv(time,recur)~genotype))

> plot(fit,lty=1:2,
+          ylab='Probability of Hemorrhage-Free Survival',
+          xlab='Time (in Months)'
+ )
> legend('bottomleft',c('Group 1', 'Group 2'),lty=1:2,bty='n')
```

As illustrated in Figure 6.2, patients that were homozygous *e3/e3* (Group 1) seem to have significantly greater survival.

```
> with(hemorrhage, survdiff(Surv(time,recur)~genotype))

Call:
survdiff(formula = Surv(time, recur) ~ genotype)

n=70, 1 observation deleted due to missingness.

             N Observed Expected (O-E)^2/E (O-E)^2/V
genotype=0 32        4     9.28      3.00      6.28
genotype=1 38       14     8.72      3.19      6.28

 Chisq= 6.3  on 1 degrees of freedom, p= 0.0122
```

Note that the log-rank test statistic is 6.3 with p-value 0.0122 based on a null χ^2-distribution with 1 degree of freedom. Thus the log-rank test confirms the difference in survival time of the two groups. ∎

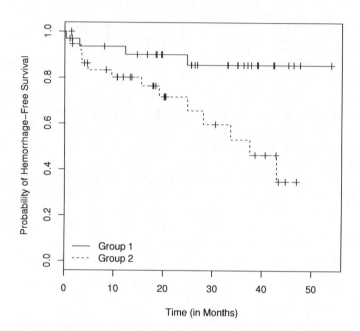

FIGURE 6.2
Plots of Kaplan–Meier estimated survival distributions.

6.2.1 Gehan's Test

Gehan's test, (see Higgins (2003)), sometimes referred to as the Gehan–Wilcoxon test, is an alternative to the log-rank test. Gehan's method is a generalization of the Wilcoxon procedure discussed in Chapter 3. Suppose in a randomized controlled trial subjects are randomized to one of two treatments, say, with survival times represented by X and Y. Represent the sample as $X_1, \ldots, X_{n_1}$ and $Y_1, \ldots, Y_{n_2}$ with a censored observation denoted by with a plus sign, X_i^+, for example. Only unambiguous pairs of observations are used. Not used are ambiguous observations such as when an observed X is greater than a censored Y $(X_i > Y_j^+)$ or when both observations are censored. The test statistic is defined as the number of times each of the X clearly beats Y minus the number of times Y clearly beats X. Let S_1 denote the set of uncensored observations, S_2 denote the set of observations for which X is censored and Y is uncensored, and S_3 denote the set where Y is censored and X is uncensored. Then Gehan's test statistic can be represented as

$$U = \left(\#_{S_1}\{X_i > Y_j\} + \#_{S_2}\{X_i^+ \geq Y_j\} \right) - \left(\#_{S_1}\{Y_j > X_i\} + \#_{S_3}\{Y_j^+ \geq X_i\} \right).$$

TABLE 6.4
Survival Times (in days) for Undergoing Standard Treatment (S) and a New Treatment (N).

S	94	180+	741	1133	1261	382	567+	988	1355+
N	155	375	951+	1198	175	521	683+	1216+	

Example 6.2.4 (Higgins' Cancer Data). Example 7.3.1 of Higgins (2003) describes an experiment to assess the effect of a new treatment relative to a standard. The data are in the dataset `cancertrt`, but for convenience the data are also given in Table 6.4. We illustrate the computation of Gehan's test based on the npsm function `gehan.test`. There are three required arguments to the function: the survival time, an indicator variable indicating that the survival time corresponds to an event (and is not censored), and a dichotomous variable representing one of two treatments; see the output of the `args` function below.

```
> args(gehan.test)

function (time, event, trt)
NULL
```

We use the function `gehan.test` next on the `cancertrt` dataset.

```
> with(cancertrt,gehan.test(time,event,trt))

statistic =  -0.6071557 , p-value =   0.5437476
```

The results agree with those in Higgins. The two-sided p-value $= 0.5437$ which is not significant. As a final note, using the `survdiff` function with `rho=1` gives the Peto–Peto modification of the Gehan test. ∎

6.3   Cox Proportional Hazards Models

As in the last section, let T denote the time until the event of an experimental unit. Let $\boldsymbol{x}$ denote the corresponding $p \times 1$ vector of covariates. Assume that T is a continuous random variable with respective pdf and cdf denoted by $f(t)$ and $F(t)$. Let $S(t) = 1 - F(t)$ denote the survival time of T. Let T_0 denote a baseline response; i.e., a response in the absence of all covariate effects.

The hazard function of T, which is often interpreted as the instantaneous chance of the event (death), is defined as

$$h(t) = \frac{f(t)}{S(t)};$$

see expression (6.7) for a formal definition. For a simple but much used example, assume that T_0 has the exponential distribution with pdf $f(t) = \lambda_0 \exp\{-\lambda_0 t\}$, $t > 0$. Then it is easy to show that the hazard function of T_0 has the constant value of λ_0. The proportional hazards model assumes that the hazard function of T is given by

$$\lambda(t; \boldsymbol{x}) = \lambda_0 e^{\boldsymbol{\beta}^T \boldsymbol{x}} \tag{6.3}$$

where $\boldsymbol{x}$ is a $p \times 1$ vector of covariates and $\boldsymbol{\beta}$ is a $p \times 1$ vector of parameters. Note that the hazard function of T is proportional to that of T_0.

To illustrate these ideas, assume that T_0 has constant hazard λ_0. Suppose the only covariate is an indicator variable w which is either 0 or 1 depending on whether a subject is not treated or treated. Assuming a proportional hazards model, the hazard function of T is given by

$$\lambda(t; w) = \lambda_0 e^{w\Delta}. \tag{6.4}$$

The hazard ratio of the experimental treatment relative to the control is then e^Δ. That is, Δ has the interpretation of log hazard; a value < 1 (less hazardous) favors the experimental treatment and a value > 1 (more hazardous) favors the control. Further examples are given in Section 7.4.2 of Cook and DeMets (2008).

The proportional hazards model developed by (Cox 1972) is a semiparametric model which does not necessarily specify the hazard function; only the relative effect of covariates is estimated. In the simple case under discussion it can be used to estimate the parameter Δ as shown in the following example.

Example 6.3.1 (Hemorrhage data example, Continued). As a first example, we again consider Example 6.2.3 concerning the hemorrhage data from the previous section. Using the function `coxph` from the `survival` package we obtain an estimate Δ and corresponding inference.

```
> fit<-coxph(Surv(time,recur)~genotype,data=hemorrhage)
> summary(fit)

Call:
coxph(formula = Surv(time, recur) ~ genotype, data = hemorrhage)

  n= 70, number of events= 18
   (1 observation deleted due to missingness)

            coef exp(coef) se(coef)     z Pr(>|z|)
genotype 1.3317    3.7874   0.5699 2.337   0.0195 *
---
Signif. codes:  0 '***' 0.001 '**' 0.01 '*' 0.05 '.' 0.1 ' ' 1

          exp(coef) exp(-coef) lower .95 upper .95
genotype      3.787      0.264     1.239     11.57
```

```
Concordance= 0.622   (se = 0.065 )
Rsquare= 0.09    (max possible= 0.851 )
Likelihood ratio test= 6.61  on 1 df,    p=0.01015
Wald test             = 5.46  on 1 df,    p=0.01946
Score (logrank) test = 6.28  on 1 df,    p=0.01219
```

From the output we see $\hat{\Delta} = 1.3317$ which indicates an increased risk for Group 2, those with heterozygous genotype. We also observe the estimated risk of hemorrhage for being heterozygous (Group 2) is 3.787 over being homozygous (Group 1). A 95% confidence interval is also given as $(1.239, 11.57)$. Notice that the value of the score test statistics is the same as from the last section. ∎

More generally, assume that the baseline hazard function is $\lambda_0(t)$. Assume that the hazard function of T is

$$\lambda(t; \boldsymbol{x}) = \lambda_0(t)e^{\boldsymbol{\beta}^T \boldsymbol{x}}.$$

Notice that the hazard ratio of two covariate patterns (e.g. for two subjects) is independent of baseline hazard

$$\frac{\lambda(t; \boldsymbol{x}_1)}{\lambda(t; \boldsymbol{x}_2)} = e^{\boldsymbol{\beta}(\boldsymbol{x}_1 - \boldsymbol{x}_2)}.$$

We close this section with the following example concerning an investigation with treatment at two levels and several covariates.

Example 6.3.2 (DES for treatment of prostate cancer). The following example is taken from Collett (2003); data are available from the publisher's website. Under investigation in this clinical trial was the pharmaceutical agent diethylstilbestrol DES; subjects were assigned treatment to 1.0 mg DES (treatment = 2) or to placebo (treatment = 1). Covariates include age, serum hemoglobin level, size, and the Gleason index.

In Exercise 6.5.2 the reader is asked to obtain the full model fit for the Cox proportional hazards model. Several of the explanatory variables are nonsignificant, though in practice one may want to include important risk factors such as age in the final model. For demonstration purposes, we have dropped **age** and **shb** from the model. As discussed in Collett (2003), the most important predictor variables are size and index.

```
> f2<-coxph(Surv(time,event=status)~as.factor(treatment)+size+index,
+ data=prostate)
> summary(f2)

Call:
coxph(formula = Surv(time, event = status) ~ as.factor(treatment) +
    size + index, data = prostate)
```

```
n= 38, number of events= 6

                         coef exp(coef) se(coef)        z Pr(>|z|)
as.factor(treatment)2 -1.11272   0.32866  1.20313 -0.925   0.3550
size                   0.08257   1.08608  0.04746  1.740   0.0819 .
index                  0.71025   2.03450  0.33791  2.102   0.0356 *
---
Signif. codes:  0 '***' 0.001 '**' 0.01 '*' 0.05 '.' 0.1 ' ' 1

                      exp(coef) exp(-coef) lower .95 upper .95
as.factor(treatment)2    0.3287     3.0426   0.03109     3.474
size                     1.0861     0.9207   0.98961     1.192
index                    2.0345     0.4915   1.04913     3.945

Concordance= 0.873   (se = 0.132 )
Rsquare= 0.304    (max possible= 0.616 )
Likelihood ratio test= 13.78  on 3 df,   p=0.003226
Wald test        = 10.29  on 3 df,   p=0.01627
Score (logrank) test = 14.9  on 3 df,   p=0.001903
```

These data suggest that the Gleason Index is a significant risk factor of mortality (p-value=0.0356). Size of tumor is marginally significant (p-value=0.0819). Given that $\hat{\Delta} = -1.11272 < 1$ it appears that DES lowers risk of mortality; however, the p-value $= 0.3550$ is nonsignificant. ∎

6.4   Accelerated Failure Time Models

In this section we consider analysis of survival data based on an accelerated failure time model. We assume that all survival times are observed. Rank-based analysis with censored survival times is considered in Jin et al. (2003).

Consider a study on experimental units (subjects) in which data are collected on the time until failure of the subjects. Hence, the setup for this section is the same as in the previous two sections of this chapter, with time until event replaced by time until failure. For such an experiment or study, let T be the time until failure of a subject and let x be the vector of associated covariates. The components of x could be indicators of an underlying experimental design and/or concomitant variables collected to help explain random variability. Note that $T > 0$ with probability one. Generally, in practice, T has a skewed distribution. As in the last section, let the random variable T_0 denote the baseline time until failure. This is the response in the absence of all covariates.

In this section, let $g(t; x)$ and $G(t; x)$ denote the pdf and cdf of T, respectively. In the last section, we introduced the hazard function $h(t)$. A more formal definition of the hazard function is the limit of the rate of instantaneous failure at time t; i.e.,

$$h(t; \boldsymbol{x}) = \lim_{\Delta t \downarrow 0} \frac{P[t < T \le t + \Delta t | T > t; \boldsymbol{x}]}{\Delta t}$$

$$= \lim_{\Delta t \downarrow 0} \frac{g(t; \boldsymbol{x}) \Delta t}{\Delta t (1 - G(t; \boldsymbol{x}))} = \frac{g(t; \boldsymbol{x})}{1 - G(t; \boldsymbol{x})}. \qquad (6.5)$$

Models frequently used with failure time data are the log-linear models

$$Y = \alpha + \boldsymbol{x}^T \boldsymbol{\beta} + \epsilon, \qquad (6.6)$$

where $Y = \log T$ and ϵ is random error with respective pdf and cdf $f(s)$ and $F(s)$. We assume that the random error ϵ is free of $\boldsymbol{x}$. Hence, the baseline response is given by $T_0 = \exp\{\epsilon\}$. Let $h_0(t)$ denote the hazard function of T_0. Because

$$T = \exp\{Y\} = \exp\{\alpha + \boldsymbol{x}^T \boldsymbol{\beta} + \epsilon\} = \exp\{\alpha + \boldsymbol{x}^T \boldsymbol{\beta}\} \exp\{\epsilon\} = \exp\{\alpha + \boldsymbol{x}^T \boldsymbol{\beta}\} T_0,$$

it follows that the hazard function of T is

$$h_T(t; \boldsymbol{x}) = \exp\{-(\alpha + \boldsymbol{x}^T \boldsymbol{\beta})\} h_0(\exp\{-(\alpha + \boldsymbol{x}^T \boldsymbol{\beta})\} t). \qquad (6.7)$$

Notice that the effect of the covariate $\boldsymbol{x}$ either accelerates or decelerates the instantaneous failure time of T; hence, log-linear models of the form (6.6) are generally called **accelerated failure time models**.

If T_0 has an exponential distribution with mean $1/\lambda_0$, then the hazard function of T simplifies to:

$$h_T(t; \boldsymbol{x}) = \lambda_0 \exp\{-(\alpha + \boldsymbol{x}^T \boldsymbol{\beta})\}; \qquad (6.8)$$

i.e., Cox's proportional hazard function given by expression (6.3) of the last section. In this case, it follows that the density function of ϵ is the extreme-valued pdf given by

$$f(s) = \lambda_0 e^s \exp\left\{-\lambda_0 e^s\right\}, \quad -\infty < s < \infty. \qquad (6.9)$$

Accelerated failure time models are discussed in Kalbfleisch and Prentice (2002). As a family of possible error distributions for ϵ, they suggest the generalized log F family; that is, $\epsilon = \log T_0$, where down to a scale parameter, T_0 has an F-distribution with $2m_1$ and $2m_2$ degrees of freedom. In this case, we say that $\epsilon = \log T_0$ has a $GF(2m_1, 2m_2)$ distribution. Kalbfleisch and Prentice discuss this family for $m_1, m_2 \ge 1$; while McKean and Sievers (1989) extended it to $m_1, m_2 > 0$. This provides a rich family of distributions. The distributions are symmetric for $m_1 = m_2$; positively skewed for $m_1 > m_2$; negatively skewed for $m_1 < m_2$; moderate to light-tailed for $m_1, m_2 > 1$; and heavy tailed for $m_1, m_2 \le 1$. For $m_1 = m_2 = 1$, ϵ has a logistic distribution, while as $m_1 = m_2 \to \infty$ the limiting distribution of ϵ is normal. Also, if one of m_i is one while the other approaches infinity, then the GF distribution approaches an extreme valued-distribution, with pdf of the form (6.9). So at least in the limit, the accelerated GF models encompass the proportional hazards models. See Kalbfleisch and Prentice (2002) and Section 3.10 of Hettmansperger and McKean (2011) for discussion.

The accelerated failure time models are linear models so the rank-based fit and associated inference using Wilcoxon scores can be used for analyses. By a prudent choice of a score function, though, this analysis can be optimized. We next discuss optimal score functions for these models and show how to compute analyses based on the them using Rfit. We begin with the proportional hazards model and then discuss the scores for the generalized log F-family.

Suppose a proportional hazard model is appropriate, where the baseline random variable T_0 has an exponential distribution with mean $1/\lambda_0$. Then ϵ has the extreme valued pdf given by (6.9). Then as shown in Exercise 6.5.5 the optimal rank-based score function is $\varphi(u) = -1 - \log(1-u)$, for $0 < u < 1$. A rank-based analysis using this score function is asymptotically fully efficient. These scores are in the package npsm under the name logrankscores. The left panel of Figure 6.3 contains a plot of these scores, while the right panel shows a graph of the corresponding extreme valued pdf, (6.9). Note that the density has very light right-tails and much heavier left-tails. To guard against the influence of large (absolute) observations from the left-tails, the scores are bounded on the left, while their behavior on the right accommodates light-tailed error structure. The scores, though, are unbounded on the right and, hence, the resulting R analysis is not bias robust. In the sensitivity analysis discussed in McKean and Sievers (1989), the R estimates based on these scores were much less sensitive to outliers than the maximum likelihood estimates. Similar to the normal scores, these log rank scores appear to be technically bias robust.

We illustrate the use of these scores in the next example.

Example 6.4.1 (Simulated Exponential Data). The data for this model are generated from a proportional hazards model with $\lambda = 1$ based on the code eps <- log(rexp(10)); x=1:10; y = round(4*x+eps,digits=2). The actual data used are given in Exercise 6.5.11. Using Rfit with the log-rank score function, we obtain the fit of this dataset:

```
> fit <- rfit(y~x,scores=mylogrank)
> summary(fit)

Call:
rfit.default(formula = y ~ x, scores = mylogrank)

Coefficients:
            Estimate Std. Error t.value   p.value
(Intercept) -1.60687    1.49251 -1.0766     0.313
x            4.19125    0.22496 18.6310 7.107e-08 ***
---
```

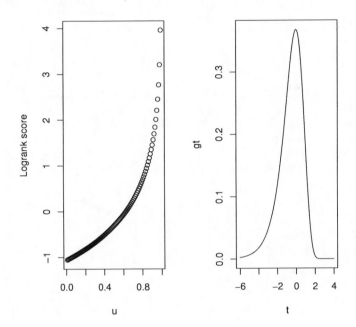

FIGURE 6.3
Log-rank score function.

Signif. codes: 0 '***' 0.001 '**' 0.01 '*' 0.05 '.' 0.1 ' ' 1

Multiple R-squared (Robust): 0.9287307
Reduction in Dispersion Test: 104.2504 p-value: 1e-05

Note that the true slope of 4 is included in the approximate 95% confidence interval $4.19 \pm 2.31 \cdot 0.22$. ∎

Next, suppose that the random errors in the accelerated failure time model, (6.6), have down to a scale parameter, a $GF(2m_1, 2m_2)$ distribution. Then as shown on page 234 of Hettmansperger and McKean (2011) the optimal score function is

$$\varphi_{m1_1, m_2}(u) = \frac{m_1 m_2 [\exp\{F^{-1}(u)\} - 1]}{m_2 + m_1 \exp\{F^{-1}(u)\}}, \quad m_1 > 0, m_2 > 0, \qquad (6.10)$$

where F is the cdf of ϵ. Note, for all values of m_1 and m_2, these score functions are bounded over the interval $(0, 1)$; hence, the corresponding R analysis is biased robust. These scores are called the generalized log-F scores (GLF). The software npsm contains the necessary R code logfscores to add these scores

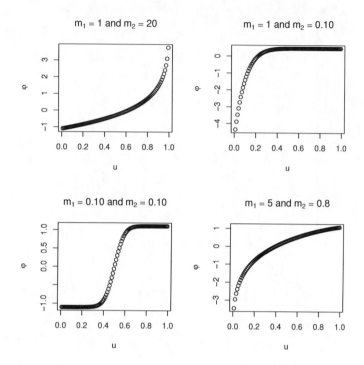

FIGURE 6.4
GLF scores for various settings of m_1 and m_2.

to the class of scores. For this code, we have used the fact that the *pth* quantile of the $F_{2m_1,2m_2}$ cdf satisfies

$$q = \exp\{F_\epsilon^{-1}(p)\} \text{ where } q = F_{2m_1,2m_2}^{-1}(p).$$

The default values are set at $m_1 = m_2 = 1$, which gives the Wilcoxon scores. Figure 6.4 shows the diversity of these scores for different values of m_1 and m_2. It contains plots of four of the scores. The upper left corner graph displays the scores for $m_1 = 1$ and $m_2 = 20$. These are suitable for error distributions which have moderately heavy (heaviness of a logistic distribution) left-tails and very light right-tails. In contrast, the scores for $m_1 = 1$ and $m_2 = 0.10$ are appropriate for moderately heavy left-tails and very heavy right-tails. The lower left panel of the figure is a score function designed for heavy tailed and symmetric distributions. The final plot, $m_1 = 5$ and $m_2 = 0.8$, are appropriate for moderate left-tails and heavy right-tails. But note from the degree of downweighting that the right-tails for this last case are clearly not as heavy as for the two cases with $m_2 = 0.10$.

The next example serves as an application of the log F-scores.

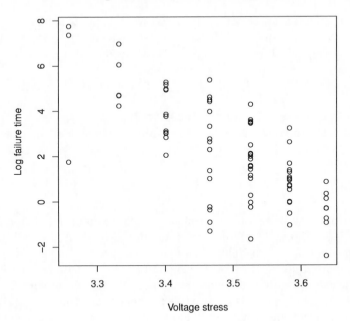

FIGURE 6.5
Log failure times of the insulation fluid versus the voltage stress.

Example 6.4.2 (Insulating Fluid Data). Hettmansperger and McKean (2011) present an example involving failure time (T) of an electrical insulating fluid subject to seven different levels of voltage stress (x). The data are in the dataset `insulation`. Figure 6.5 shows a scatterplot of the log of failure time $(Y = \log T)$ versus the voltage stress. As voltage stress increases, time until failure of the insulating fluid decreases. It appears that a simple linear model suffices. In their discussion, Hettmansperger and McKean recommend a rank-based fit based on generalized log F-scores with $m_1 = 1$ and $m_2 = 5$. This corresponds to a distribution with left-tails as heavy as a logistic distribution and right-tails lighter than a logistic distribution; i.e., moderately skewed left. The following code-segment illustrates computation of the rank-based fit of these data based on this log F-score.

```
> myscores <- logfscores
> myscores@param=c(1,5)
> fit <- rfit(logfail~voltstress,scores=myscores)
> summary(fit)
```

Call:

```
rfit.default(formula = logfail ~ voltstress, scores = myscores)
```

```
Coefficients:
            Estimate Std. Error t.value   p.value
(Intercept)  63.9596     6.5298   9.795 5.324e-15 ***
voltstress  -17.6624     1.8669  -9.461 2.252e-14 ***
---
Signif. codes:  0 '***' 0.001 '**' 0.01 '*' 0.05 '.' 0.1 ' ' 1
```

```
Multiple R-squared (Robust): 0.5092232
Reduction in Dispersion Test: 76.78138 p-value: 0
```

```
> fit$tauhat
```

```
[1] 1.572306
```

Not surprisingly, the estimate of the slope is highly significant. As a check on goodness-of-fit, Figure 6.6 presents the Studentized residual plot and the $q-q$ plot of the Studentized residuals versus the quantiles of a log F-distribution with the appropriate degrees of freedom 2 and 10. For the $q-q$ plot, the population quantiles are the quantiles of a log f-distribution with 2 and 10 degrees of freedom. This plot is fairly linear, indicating[1] that an appropriate choice of scores was made. The residual plot indicates a good fit. The outliers on the left are mild and, based on the $q-q$ plot, follow the pattern of the log F-distribution with 2 and 10 degrees of freedom. ∎

6.5   Exercises

6.5.1. Using the data discussed in Example 6.2.4:

(a) Obtain a plot of the Kaplan–Meier estimates for the two treatment groups.

(b) Obtain the p-value based on the log-rank statistic.

(c) Obtain the p-value based on the Peto–Peto modification of the Gehan statistic.

6.5.2. Obtain full model fit of the prostate cancer data discussed in Example 6.3.2. Include age, serum haemoglobin level, size, and Gleason index. Comment on the similarity or dissimilarity of the estimated regression coefficients to those obtained in Example 6.3.2.

---

[1]See the discussion in Section 3.10 of Hettmansperger and McKean (2011).

Studentized residual plot

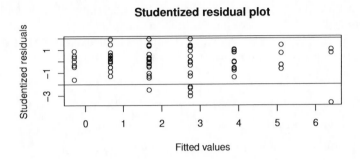

Studentized q–q plot

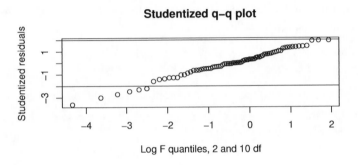

FIGURE 6.6
The top panel contains the Studentized residual plot of the rank-based fit using generalized log F-scores with 2 and 10 degrees of freedom. The bottom panel shows the $q-q$ plot of Studentized residuals versus log F-scores with 2 and 10 degrees of freedom.

6.5.3. For the dataset `hodgkins`, plot Kaplan–Meier estimated survival curves for both treatments. Note treatment code 1 denotes radiation of affected node and treatment code 2 denotes total nodal radiation.

6.5.4. To simulate survival data, often it is useful to simulate multiple time points. For example the time to event and the time to end of study. Then, events occurring after the time to end of study are censored. Suppose the time to event of interest follows an exponential distribution with mean 5 years and the time to end of study follows an exponential distribution with a mean of 1.8 years. For a sample size $n = 100$ simulate survival times from this model. Plot the Kaplan–Meier estimate.

6.5.5. Show that the optimal rank-based score function is $\varphi(u) = -1 - \log(1 - u)$, for $0 < u < 1$ for random variables which have an extreme valued distribution (6.9). In this case, the generated scores are called the log-rank scores

6.5.6. Consider the dataset `rs`. This is simulated data from a simple regression model with the true slope parameter at 0.5. The first column is the independent variable x while the second column is the dependent variable y. Obtain the following three fits of the model: least squares, Wilcoxon rank-based, and rank-based using `logfscores` with $m_1 = 1$ and $m_2 = 0.10$.

    (a) Scatterplot the data and overlay the three fits.

    (b) Obtain Studentized residual plots of all three fits.

    (c) Based on Parts (a) and (b) which fit is worst?

    (d) Compare the two rank-based fits in terms of precision (estimates of τ_φ). Which fit is better?

6.5.7. Generate data from a linear model with log-F errors with degrees of freedom 4 and 8 using the following code

```
n <- 75; m1 <- 2; m2 <- 4; x<-rnorm(n,50,10)
errs1 <- log(rf(n,2*m1,2*m2)); y1 <- x + 30*errs1
```

    (a) Using `logfscores`, obtain the optimal scores for this dataset.

    (b) Obtain side-by-side plots of the pdf of the random errors and the scores. Comment on the plot.

    (c) Fit the simple linear model for this data using the optimal scores. Obtain a residual analysis including a Studentized residual plot and a normal q–q plot. Comment on the plots and the quality of the fit.

    (d) Obtain a histogram of the residuals for the fit in part (b). Overlay the histogram with an estimate of the density and compare it to the plot of the pdf in part (a).

    (e) Obtain a summary of the fit of the simple linear model for this data using the optimal scores. Obtain a 95% confidence interval for the slope parameter β. Did the interval trap the true parameter?

    (f) Use the fit to obtain a confidence interval for the expected value of y when $x = 60$.

6.5.8. For the situation described in Exercise 6.5.7, obtain a simulation study comparing the mean squared errors of the estimates of slope using fits based on Wilcoxon scores and the optimal scores. Use 10,000 simulations.

6.5.9. Consider the failure time data discussed in Example 6.4.2. Recall that the generalized log F-scores with $2m_1 = 2$ and $2m_2 = 10$ degrees of freedom were used to compute the rank-based fit. The Studentized residuals from this fit were then used in a q–q plot to check goodness-of-fit based on the strength of linearity in the plot, where the population quantiles were obtained from a log F-distribution with 2 and 10 degrees of freedom. Obtain the rank-based fits based on the Wilcoxon scores, normal scores, and log F-scores with $2m_1 = 10$

and $2m_2 = 2$. For each, obtain the q–q plot of Studentized residuals using as population quantiles the normal distribution, the logistic distribution, and the log F-distribution with 10 and 2 degrees of freedom, respectively. Compare the plots. Which, if any, is most linear?

6.5.10. Suppose we are investigating the relationship between a response Y and an independent variable x. In a planned experiment, we record responses at r values of x, $x_1 < x_2 < \cdots < x_r$. Suppose n_i independent replicates are obtained at x_i. Let Y_{ij} denote the response for the jth replicate at x_i. Then the model for a linear relationship is

$$Y_{ij} = \alpha + x_{ij}\beta + e_{ij}, \quad i = 1, \ldots, r; j = 1, \ldots, n_i. \tag{6.11}$$

In this setting, we can obtain a **lack-of-fit** test. For this test, the null hypothesis is Model (6.11). For the alternative, we take the most general model which is a one-way design with r groups; i.e., the model

$$Y_{ij} = \mu_i + e_{ij}, \quad i = 1, \ldots, r; j = 1, \ldots, n_i, \tag{6.12}$$

where μ_i is the median (or mean) of the ith group (responses at x_i). The rank-based drop in dispersion is easily formulated to test these hypotheses. Select a score function φ. Let $D(\text{RED})$ denote the minimum value of the dispersion function when Model (6.11) is fit and let $D(\text{FULL})$ denote the minimum value of the dispersion function when Model (6.12) is fit. The F_φ test statistic is

$$F_\varphi = \frac{[D(\text{RED}) - D(\text{FULL})]/(r - 2)}{\hat{\tau}_\varphi}.$$

This test statistic should be compared with F-critical values having $r - 2$ and $n - r$ degrees of freedom, where $n = \sum_i n_i$ is the total sample size. In general the drop in dispersion test is computed by the function `drop.test`. Carry out this test for the data in Example 6.4.2 using the log F-scores with $2m_1 = 2$ and $2m_2 = 10$ degrees of freedom.

6.5.11. The data for Example 6.4.1 are:

x	1	2	3	4	5	6	7	8	9	10
y	2.84	6.52	6.87	16.43	18.17	25.24	28.15	31.65	36.37	38.84

(a) Using `Rfit`, verify the analysis presented in Example 6.4.1.

(b) Obtain Studentized residuals from the fit. Comment on the residual plot.

(c) Obtain the q–q plot of the sorted residuals of Part (b) versus the quantiles of the random variable ε which is distributed as the log of an exponential. Comment on linearity in the q–q plot.

7

Regression II

7.1 Introduction

In Chapter 4 we introduced rank-based fitting of linear models using Rfit. In this chapter, we discuss further topics for rank-based regression. These include high breakdown fits, diagnostic procedures, weighted regression, nonlinear models, and autoregressive time series models. We also discuss optimal scores for a family of skew normal distributions and present an adaptive procedure for regression estimation based on a family of Winsorized Wilcoxon scores.

Let $Y = [y_1, \ldots, y_n]^T$ denote an $n \times 1$ vector of responses. Then the matrix version of the linear model, (4.2), is

$$Y = \alpha 1 + X\beta + e \tag{7.1}$$

where $X = [x_1, \ldots, x_n]^T$ is an $n \times p$ design matrix, and $e = [e_1, \ldots, e_n]^T$ is an $n \times 1$ vector of error terms. Assume for discussion that $f(t)$ and $F(t)$ are the pdf and cdf of e_i, respectively. Assumptions differ for the various sectional topics.

Recall from expression (4.10) that the rank-based estimator $\hat{\beta}_\varphi$ is the vector that minimizes the rank-based distance between Y and $X\beta$; i.e., $\hat{\beta}_\varphi$ is defined as

$$\hat{\beta}_\varphi = \text{Argmin}\|y - X\beta\|_\varphi, \tag{7.2}$$

where the norm is defined by

$$\|v\|_\varphi = \sum_{i=1}^{n} a[R(y_i - x_i^T \beta)](y_i - x_i^T \beta), \quad v \in R^n, \tag{7.3}$$

and the scores $a(i) = \varphi[i/(n+1)]$ for a specified score function $\varphi(u)$ defined on the interval $(0, 1)$ and satisfying the standardizing conditions given in (3.12).

Note that the norm is invariant to the intercept parameter; but, once β is estimated, the intercept α is estimated by the median of the residuals. That is,

$$\hat{\alpha} = \text{med}_i\{y_i - x_i^T \hat{\beta}_\varphi\}. \tag{7.4}$$

The rank-based residuals are defined by

$$\hat{e}_i = y_i - \hat{\alpha} - x_i^T \hat{\beta}_\varphi, \quad i = 1, 2, \ldots, n. \tag{7.5}$$

Recall that the joint asymptotic distribution of the rank-based estimates is multivariate normal with the covariance structure as given in (4.12).

As discussed in Chapter 3, the rank-based estimates are generally highly efficient estimates. Further, as long as the score function is bounded, the influence function of $\hat{\beta}_\varphi$ is bounded in the Y-space (response space). As with LS estimates, though, the influence function is unbounded in the x-space (factor space). In the next section, we present a rank-based estimate which has bounded influence in both spaces and which can attain the maximal 50% breakdown point.

7.2 High Breakdown Rank-Based Fits

High breakdown rank-based (HBR) estimates were developed by Chang et al. (1999) and are fully discussed in Section 3.12 of Hettmansperger and McKean (2011). To obtain HBR fits of linear models, a suite of R functions (ww) was developed by Terpstra and McKean (2005). We use a modified version, hbrfit, of ww to compute HBR fits.[1]

The objective function for HBR estimation is a weighted Wilcoxon dispersion function given by

$$\|\boldsymbol{v}\|_{HBR} = \sum_{i<j} b_{ij}|v_i - v_j| \qquad (7.6)$$

where $b_{ij} \geq 0$ and $b_{ij} = b_{ji}$. The HBR estimator of $\boldsymbol{\beta}$ minimizes this objective function, which we denote by

$$\hat{\boldsymbol{\beta}}_{HBR} = \text{Argmin}\|\boldsymbol{y} - \boldsymbol{X\beta}\|_{HBR}. \qquad (7.7)$$

As with the rank-based estimates the intercept α is estimated as the median of the residuals; that is,

$$\hat{\alpha} = \text{med}_i\{y_i - \boldsymbol{x}^T\hat{\boldsymbol{\beta}}_{HBR}\}. \qquad (7.8)$$

As shown in Chapter 3 of Hettmansperger and McKean (2011), if all the weights are one (i.e., $b_{ij} \equiv 1$) then $\| \cdot \|_{HBR}$ is the Wilcoxon norm. Thus the question is, what weights should be chosen to yield estimates which are robust to outliers in both the x- and y-spaces? In Section 7.2.1, we discuss the HBR weights implemented in hbrfit which achieve 50% breakdown. For now, though, we illustrate their use and computation with several examples.

Stars Data

In this subsection we present an example to illustrate the usage of the weighted Wilcoxon code hbrfit to compute HBR estimates. This example uses the

---

[1] See https://github.com/kloke/book for more information.

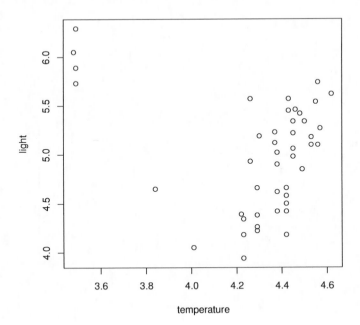

FIGURE 7.1
Scatterplot of stars data.

`stars` dataset which is from Rousseeuw et al. (1987). The data are from an astronomy study on the star cluster CYG OB1. The cluster contains 47 stars. Measurements were taken on light intensity and temperature. The response variable is log light intensity and the explanatory variable is log temperature. As is apparent in the scatterplot displayed in Figure 7.1 there are several outliers: there are four stars with lower temperature and higher light intensity than the other members of the cluster. These four stars are labeled giant stars in this dataset. The others are labeled main sequence stars, except for the two with log temperature 3.84 and 4.01 which are between the giant and main sequence stars.

In Figure 7.2, the Wilcoxon (WIL), high breakdown (HBR), and least squares (LS) fits are overlaid on the scatterplot. As seen in Figure 7.2, both the least squares and Wilcoxon fit are affected substantially by the outliers; the HBR fit, however, is robust.

The HBR fit is computed as

```
> fitHBR<-hbrfit(stars$light ~ stars$temperature)
```

As we have emphasized throughout the book the use of residuals, in particular

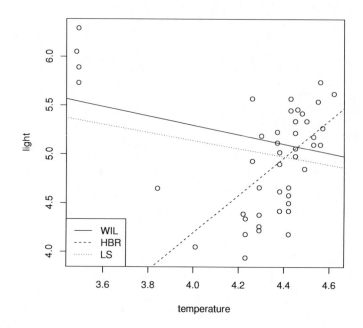

FIGURE 7.2
Scatterplot of stars data with fitted regression lines overlaid.

Studentized residuals, are essential to the model building process. Studentized residuals are available through the command **rstudent**. In addition, a set of diagnostic plots can be obtained using **diagplot**. For HBR fit of the stars data, Figure 7.3 displays these diagnostic plots, which resulted from the code:

```
> diagplot(fitHBR)
```

Note from these plots in Figure 7.3 that the Studentized residuals of the HBR fit clearly identify the 4 giant stars. They also identify the two stars between the giant and main sequence stars.

Finally, we may examine the estimated regression coefficients and their standard errors in the table of regression coefficients with the command **summary**; i.e.,

```
> summary(fitHBR)
```

```
Call:
hbrfit(formula = stars$light ~ stars$temperature)
```

```
Coefficients:
```

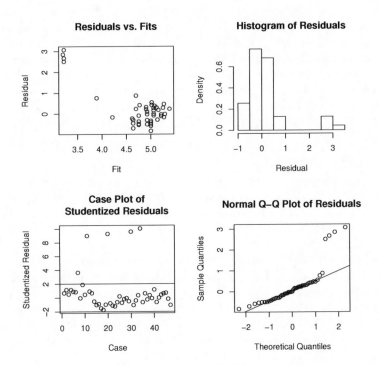

FIGURE 7.3
Diagnostic plots based on the HBR fit of the stars data.

```
                  Estimate Std. Error t.value  p.value
(Intercept)       -3.46917    1.64733 -2.1059  0.04082 *
stars$temperature  1.91667    0.38144  5.0248 8.47e-06 ***
---
Signif. codes:  0 '***' 0.001 '**' 0.01 '*' 0.05 '.' 0.1 ' ' 1

Wald Test: 25.24853 p-value: 1e-05
```

The estimate of intercept is -3.47 (se $= 1.65$). The estimate of slope is 1.92 (se $= 0.38$). Critical values based on a t-distribution with $n - p - 1$ degrees of freedom are recommend for inference; for example p-values in the coefficients table for the stars data are based on a t_{45} distribution. Also displayed is a Wald test of $H_0 : \boldsymbol{\beta} = \mathbf{0}$.

7.2.1 Weights for the HBR Fit

Let $\boldsymbol{X}_c$ be the centered design matrix. For weights it seems reasonable to downweight points far from the center of the data. The traditional distances are the leverage values $h_i = n^{-1} + x'_{ci}(X'_c X_c)^{-1} x_{ci}$, where x_{ci} is the vector of the *ith* row of X_c. Because the leverage values are based on the LS variance-covariance scatter matrix, they are not robust. The weights for the HBR estimate make use of the high breakdown minimum covariance determinant, **MCD**, which is an ellipsoid in p-space that covers about half of the data and yet has minimum determinant. Rousseeuw and Van Driessen (1999) present a fast computational algorithm for it. Let V and v_c denote respectively the MCD and the center of the MCD. The robust distances and weights are respectively $v_{ni} = (x_{ci} - v_c)' V^{-1} (x_{ci} - v_c)$ and $w_i = \min \left\{ 1, \frac{c}{v_{ni}} \right\}$, where c is usually set at the 95*th* percentile of the $\chi^2(p)$ distribution. Note that "good" points generally have weight 1. The estimator $\widehat{\beta}^*$ (7.7) of β obtained with these weights is called a generalized R (GR) estimator. In general, this GR estimator has a bounded influence function in both the Y and the **x**-spaces and a positive breakdown. It can be computed using the suite of R functions ww with wts = "GR".

Note that the GR estimate downweights "good" points as well as "bad" points of high leverage. Due to this indiscriminate downweighting the GR estimator is less efficient than the Wilcoxon estimator. At times, the loss in efficiency can be severe. The HBR weights also use the MCD to determine weights in the **x**-space. Unlike the GR weights, though, residual information from the Y-space is also used. These residuals are based on the the **least trim squares** (LTS) estimate which is $\text{Argmin} \sum_{i=1}^h [Y - \alpha - x'\beta]^2_{(i)}$ where $h = [n/2]+1$ and (i) denotes the *ith* ordered residual. This is a high breakdown initial estimate; see Rousseeuw and Van Driessen (1999). Let $\widehat{e}_0$ denote the residuals from this initial fit.

Define the function $\psi(t)$ by $\psi(t) = 1$, t, or -1 according as $t \geq 1$, $-1 < t < 1$, or $t \leq -1$. Let σ be estimated by the initial scaling estimate $\text{MAD} = 1.483 \, \text{med}_i |\widehat{e}_i^{(0)} - \text{med}_j \{ \widehat{e}_j^{(0)} \}|$. Letting $Q_i = (x_i - v_c)' V^{-1} (x_i - v_c)$, define

$$m_i = \psi \left(\frac{b}{Q_i} \right) = \min \left\{ 1, \frac{b}{Q_i} \right\}.$$

Consider the weights

$$\widehat{b}_{ij} = \min \left\{ 1, \frac{c\hat{\sigma}}{|\hat{e}_i^{(0)}|} \frac{\hat{\sigma}}{|\hat{e}_j^{(0)}|} \min \left\{ 1, \frac{b}{\hat{Q}_i} \right\} \min \left\{ 1, \frac{b}{\hat{Q}_j} \right\} \right\}, \qquad (7.9)$$

where b and c are tuning constants. Following Chang et al. (1999), b is set at the upper $\chi^2_{.05}(p)$ quantile and c is set as

$$c = [\text{med}\{a_i\} + 3MAD\{a_i\}]^2,$$

where $a_i = \hat{e}_i^{(0)}/(MAD \cdot Q_i)$. From this point of view, it is clear that these weights downweight both outlying points in factor space and outlying responses. Note that the initial residual information is a multiplicative factor in the weight function. Hence, a good leverage point will generally have a small (in absolute value) initial residual which will offset its distance in factor space. These are the weights used for the HBR fit computed by `hbrfit`.

In general, the HBR estimator has a 50% breakdown point, provided the initial estimates used in forming the weights also have a 50% breakdown point. Further, its influence function is a bounded function in both the Y and the x-spaces, is continuous everywhere, and converges to zero as $(\mathbf{x}^*, Y^*)$ get large in any direction. The asymptotic distribution of $\widehat{\boldsymbol{\beta}}_{HBR}$ is asymptotically normal. As with all high breakdown estimates, $\widehat{\boldsymbol{\beta}}_{HBR}$ is less efficient than the Wilcoxon estimates but it regains some of the efficiency loss of the GR estimate. See Section 3.12 of Hettmansperger and McKean (2011) for discussion.

7.3 Robust Diagnostics

Diagnostics are an essential part of any analysis. The assumption of a model is a very strong statement and should not be taken lightly. As we have stressed throughout the book, diagnostic checks should be made to confirm the adequacy of the model and check the quality of fit. In this section, we explore additional diagnostics based on both highly efficient and high breakdown robust fits. These diagnostics are primarily concerned with the determination of highly influential points on the fit.

For motivation, we consider a simple dataset with two predictors and $n = 30$ data points. The scatterplot of the columns of the two-dimensional $(p = 2)$ design matrix, $\boldsymbol{X}$, is shown in Figure 7.4. The values of the x's are drawn from uniform distributions. The design matrix and observations are in the dataset `diagdata`.

Consider the four points in the upper-right corner of the plot, which are the $27th$ through $30th$ data points. As the following two sets of responses show, these points are potentially influential points on fits. The first set of responses is drawn from the model

$$Y_i = 5x_{i1} + 5x_{i2} + e_i, \tag{7.10}$$

where $e_1, \ldots, e_n$ were drawn independently from a $N(0, 1)$ distribution. We label this the "good" dataset. For this set the responses for cases 27 through 30 are respectively 7.001, 7.397, 9.191, and 8.269), which follow the model. To form the second set of responses, we negated these four responses; i.e., the observations for cases 27 though 30 are respectively $-7.001, -7.397, -9.191$, and -8.269, which of course do not follow the model. We label this second set, the "bad" dataset. We obtain the LS, Wilcoxon, and HBR fits of the two models, summarizing them in Table 7.1.

On the good dataset all three fits agree. On the bad dataset both the LS

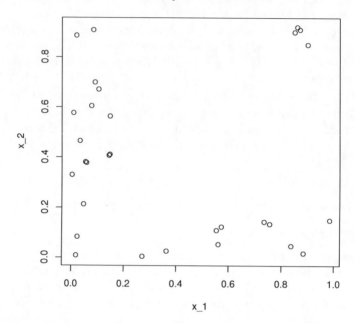

FIGURE 7.4
Scatterplot of the columns of the design matrix for the simple example.

and Wilcoxon fits were impaired while the HBR fit exhibited robustness. Thus the $27th$ through $30th$ data points are potentially influential points.

Note first that traditional diagnostic procedures are generally not efficient at these tasks. Their geometry is based on the Euclidean norm which is sensitive to outlying points in the x-space. For example, for this simple dataset, the vector of column means of the design matrix is $(0.369, 0.399)$, while the

TABLE 7.1
Estimates of the Regression for the Simple
Datasets (good and bad).

	Intercept	X1	X2
LS Good Data	0.12	4.45	4.74
Wil Good Data	0.10	4.42	4.64
HBR Good Data	-0.11	4.62	4.91
LS Bad Data	5.56	-4.49	-5.96
Wil Bad Data	6.07	-4.59	-5.83
HBR Bad Data	-0.11	4.55	4.79

vector of column medians is $(0.148, 0.38)$. Hence, the outliers have influenced the center of design based on the mean. The leverage values for the four outlying points are respectively 0.192, 0.193, 0.183, and 0.181. Note that these are less than the usual benchmark for points of high leverage which is given by $2(p+1)/n = 0.2$. Thus, if the leverage rule is followed strictly, these four points would not have been identified.

Traditional delete-one diagnostics are often used to check the quality of the fit. For example, consider the diagnostic DFFITS$_i$ which is the standardized change in the LS fitted value for case i when case i is deleted; see Belsley et al. (1980). For datasets containing a cluster of outliers in the X-space, though, when one of the cases in the cluster is deleted there are still the remaining cases in the cluster which will impair the LS fit in the same way as in the LS fit based on all the data. Hence, DFFITS$_i$ will generally be small and the points will not be detected. For the simple dataset, the values of DFFITS$_{27}$ through DFFITS$_{30}$ are respectively $-0.517, -0.583, -0.9$, and -0.742. Since the benchmark is $2\sqrt{(p+1)/n} = 0.632$, Cases 27 and 28 are not detected as influential points. The other two cases are detected but they are borderline. For example, Case 7, which is not a point of concern, has the DFFITS value of 0.70. Hence, for this example, the diagnostic DFFITS$_i$ has not been that successful.

Note that for this simple example, the HBR estimates remain essentially the same for both the "good" and "bad" datasets. We now present diagnostics based on the HBR estimates and the robust distances and LTS residuals that are used to form the HBR weights. These diagnostics are robust and are generally successful in detecting influential cases and in detecting differences between highly efficient and high breakdown robust fits.

7.3.1   Graphics

In general, consider a linear model of the form (7.1). In Section 7.2, we defined the robust distances $\sqrt{Q_i}$, where $Q_i = (x_i - v_c)'V^{-1}(x_i - v_c)$, $i = 1, \ldots, n$, V is the minimum covariance determinant (MCD), and v_c is the center of the ellipsoid V. Recall that these were used to obtain the weights in the HBR fit, see expression (7.9). Another part of these weights utilizes the standardized residuals based on the LTS fit. Rousseeuw and van Zomeren (1990) proposed as a diagnostic, the plot of these standardized residuals versus the robust distances. For the simple example with the set of bad responses, this plot is found in Figure 7.5. Note that the 4 influential cases are clearly separated from the other cases. Hence, for this example, the diagnostic plot was successful. The next segment of R code obtains the robust distances, standardized LTS residuals, and the diagnostic plot shown in Figure 7.5. The observations for the second set of data (bad data) are in the R vector ybad while the design matrix is in the R matrix x. Some caution is necessary here, because standardized residuals are not corrected for locations of the residuals in the X-space as Studentized residuals are.

Standardized LTS Residuals vs. Robust Distances

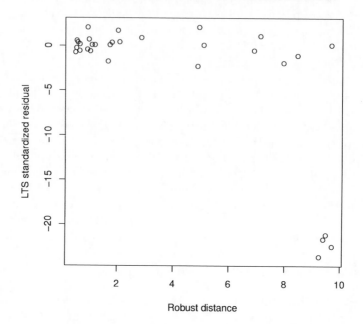

FIGURE 7.5

Standardized LTS residuals versus robust distances for the simple example with bad responses.

```
> rsdfitlts <- ltsreg(x,ybad)$resid
> srsd <- rsdfitlts /mad(rsdfitlts )
> rdis <- sqrt(robdistwts(x,ybad)$robdis2)
> plot(rdis,srsd,xlab="Robust distance",
+ ylab="LTS standardized residual")
> title(main="Standardized LTS Residuals vs. Robust Distances")
```

7.3.2 Procedures for Differentiating between Robust Fits

Recall that for the simple dataset with bad responses the differences between the Wilcoxon and HBR fits are readily apparent; see Table 7.1. We next discuss a set of formal diagnostics based on the difference between fits; see[2] McKean et al. (1996a). Consider a general linear model, say Model (7.1). The difference between estimates includes the intercept, so, for this section, let $\boldsymbol{b}^T = (\alpha, \boldsymbol{\beta}^T)$ denote the combined parameters. Then the difference between the HBR and

---

[2]See, also McKean and Sheather (2009).

Wilcoxon regression estimates is the vector $\widehat{\boldsymbol{b}}_D = \widehat{\boldsymbol{b}}_W - \widehat{\boldsymbol{b}}_{HBR}$. An effective standardization is the estimate of the variance-covariance of $\widehat{\boldsymbol{b}}_W$. A statistic which measures the total difference between the fits is

$$\text{TDBETAS} = \widehat{\boldsymbol{b}}_D^T \widehat{\boldsymbol{A}}_W^{-1} \widehat{\boldsymbol{b}}_D, \tag{7.11}$$

where $\boldsymbol{A}_W$ is the asymptotic Wilcoxon covariance matrix for linear models. Large values of TDBETAS indicate a discrepancy between the fits. A useful cutoff value is $(4(p+1)^2)/n$.

If $TDBETAS_R$ exceeds its benchmark then usually we want to determine the individual cases causing the discrepancy between the fits. Let $\widehat{y}_{W,i}$ and $\widehat{y}_{HBR,i}$ denote the respective Wilcoxon and HBR fits for the ith case. A Studentized statistic which detects the observations that differ in fit is

$$CFITS_i = (\widehat{y}_{R,i} - \widehat{y}_{HBR,i})/\sqrt{n^{-1}\widehat{\tau}_S^2 + h_{c,i}\widehat{\tau}^2}. \tag{7.12}$$

An effective benchmark for $CFITS_i$ is $2\sqrt{(p+1)/n}$. Note that the standardization of $CFITS_i$ accounts for the location of the ith case in the $\boldsymbol{X}$-space.

The objective of the diagnostic CFITS is *not* outlier deletion. Rather the intent is to identify the *few critical* data points for closer study, because these points are causing discrepancies between the highly efficient and high breakdown fits of the data. In this regard, the proposed benchmarks are meant as a heuristic aid, not a boundary to some formal critical region.

In the same way, the difference between the LS fit and either the Wilcoxon or HBR fits can be investigated. In general, though, we are interested in the difference between a highly efficient robust fit and a high breakdown robust fit. In all comparison cases, the standardization of the diagnostics is with the variance-covariance matrix of the Wilcoxon fit. For computation, the function `fitdiag`, in the collection `hbrfit`, computes these diagnostics for the Wilcoxon, HBR, GR, LS, and LTS fits. Its argument `est` specifies the difference to compute; for example, if `est=c("WIL","HBR")` then the diagnostics between the Wilcoxon and HBR fits are computed, while `est=c("LTS","WIL")` computes the diagnostics between the Wilcoxon and LTS fits. Besides the diagnostics, the associated benchmarks are returned.

From the computation, the value of TDBETAS is 43.93 which far exceeds the benchmark of 1.2; hence, the diagnostic has been successful. Even more importantly, though, is that the diagnostic CFITS in Figure 7.6 clearly flags the four influential cases. These are the points at the bottom right corner of the plot. In reading plots, such as the CFITS plot, the large gaps are important. In this case, the four influential points clearly stand out and are the ones to investigate first. As the reader is asked to show, Exercise 7.9.5, for this dataset with the set of good responses, TDBETAS is less than its benchmark and none of the CFITS values, in absolute value, exceed their benchmark.

Example 7.3.1 (Fit Diagnostics for Stars Data). The next code segment computes the robust distances and the diagnostics for the difference between the Wilcoxon and HBR fits of the stars data, discussed in Section 7.2.

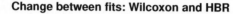

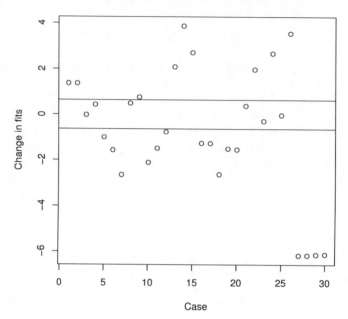

FIGURE 7.6
Plot of the changes in fits (CFITS) between the Wilcoxon and HBR fits for
the simple dataset with the set of bad responses. The horizontal lines are set
at the benchmark.

```
> dwilhbr = fitdiag(stars$temperature,stars$light,est=c("WIL","HBR"))
> tdbetas <- round(c(dwilhbr$tdbeta,dwilhbr$bmtd),digits=2)
> rdis <- sqrt(robdistwts(stars$temperature,stars$light)$robdis2)
```

The diagnostic TDBETAS has the value 67.92 which greatly exceeds its
benchmark of 0.34. Thus numerically indicating that the HBR and Wilcoxon
fits differ. The CFITS plot, right panel of Figure 7.7, clearly shows the four
giant stars (Cases 11, 20, 30, and 34). It also finds the two stars between
the giant stars and the main sequence stars, namely Cases 7 and 14. The
robust distance plot conveys similar information. For the record, the diagnostic
TDBETAS for the difference in fits between the Wilcoxon and LTS fits is
265.39, which far exceeds the benchmark. ∎

The following generated dataset illustrates the curvature problem for high
breakdown fits.

Example 7.3.2 (Curvature Data). Hettmansperger and McKean (2011),
page 267, consider a simulated quadratic model with $N(0,1)$ random errors

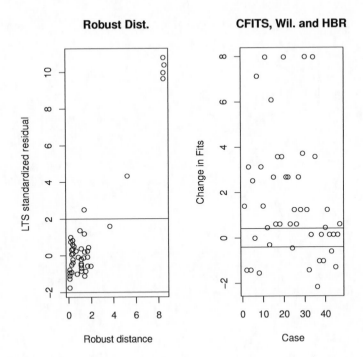

FIGURE 7.7
Robust distance plots and CFITS plot between HBR and Wilcoxon fits, stars
data.

and absolute contaminated normal xs. The model is $Y = 5.5|x| - 0.6x^2 + e$. The
scatterplot of the data overlaid with the Wilcoxon and HBR fits are shown in
Figure 7.8. The Wilcoxon and HBR estimates are:

```
> summary(fitwil)

Call:
rfit.default(formula = y ~ xmat)

Coefficients:
            Estimate Std. Error  t.value p.value
(Intercept) -0.665013  0.421598  -1.5774  0.1232
xmatx        5.946872  0.326518  18.2130  <2e-16 ***
xmat        -0.652514  0.045484 -14.3462  <2e-16 ***
---
Signif. codes:  0 '***' 0.001 '**' 0.01 '*' 0.05 '.' 0.1 ' ' 1
```

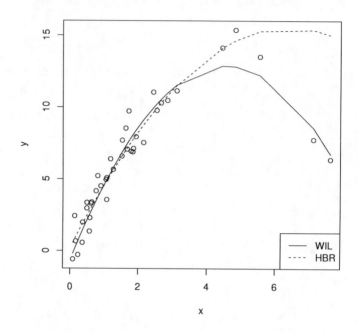

FIGURE 7.8
Scatterplot of quadratic data overlaid with fits.

```
Multiple R-squared (Robust): 0.842278
Reduction in Dispersion Test: 98.795 p-value: 0

> summary(fithbr)

Call:
hbrfit(formula = y ~ xmat)

Coefficients:
              Estimate Std. Error t.value   p.value
(Intercept)  0.175941   0.371664   0.4734 0.6387183
xmatx        4.796202   0.463747 10.3423 1.819e-12 ***
xmat        -0.373726   0.098066 -3.8110 0.0005065 ***
---
Signif. codes:  0 '***' 0.001 '**' 0.01 '*' 0.05 '.' 0.1 ' ' 1

Wald Test: 220.5558 p-value: 0
```

As the summary of coefficients and the scatterplot show the Wilcoxon and

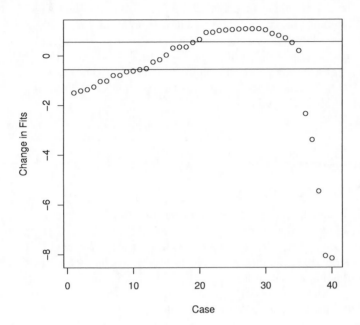

FIGURE 7.9
Plot of CFITS between HBR and Wilcoxon fits, quadratic data.

HBR fits differ. The Wilcoxon fit follows the model while the HBR fit did not detect the curvature as well as the Wilcoxon fit. For the Wilcoxon and HBR fits, the value of TDBETAS is 67.637 with benchmark 0.9; so the diagnostics on the difference of the two fits agrees with this assessment. Figure 7.9 shows the corresponding plot of the diagnostic CFITS.

The points corresponding to the largest absolute change in CFITS are at the region of most curvature in the quadratic model. ∎

As in the last example, high breakdown fits may be impaired, if curvature occurs "far" from the robust center of the data; see McKean et al. (1994) for a study on this concern for polynomial models. In the last example, the diagnostic CFITS did pinpoint this region.

7.3.3 Concluding Remarks

As the examples in this and the last sections show, in the case of messy datasets, influential points in the x-space can have a negative impact on highly efficient robust fits as well as LS fits. Hence, the use of diagnostics is recommended for detecting influential points and measuring their effect on highly efficient robust fits. In practice, if influential points are detected and found to have a negative impact on a highly efficient robust fit then we recommend using the inference based on high breakdown robust fits, such as the HBR fit. Usually, influential points in the x-space are signs of a poor design. For observational data, though, this often cannot be helped. Note, also, that such points can be part of a planned experiment such as experiments designed to detect curvature. The diagnostics discussed in this section are helpful for assessing the better fit for such cases.

7.4   Weighted Regression

For a linear model, weighted regression can be used, for example, when variances are heteroscedastic or for problems that can be solved by iterated reweighted procedures (least squares or rank-based). We begin with a general discussion that includes a simple R function which computes weighted rank-based estimates using `Rfit` .

Consider the linear model with Y as an $n \times 1$ vector of responses and X as the $n \times p$ design matrix. Write the model as

$$Y = \alpha 1 + X\beta + e = X_1 b + e, \qquad (7.13)$$

where $X_1 = [1 \ X]$ and $b = (\alpha, \beta')'$. We have placed the subscript 1 on the matrix X_1 because it contains a column of ones (intercept). Let W be an $n \times n$ matrix of weights. For example, if the covariance matrix of e is Σ, then the weight matrix that yields homogeneous variances is $W = \Sigma^{-1/2}$, where $\Sigma^{-1/2}$ is the usual square root of a positive definite matrix, (of course, in practice W needs to be estimated). The weighted model is:

$$WY = WX_1 b + We.$$

For easier notation, let $Y^* = WY$, $X^* = WX_1$, and $e^* = We$. The matrix X^* does not have a subscript 1 because generally the weights eliminate the intercept. Then we can write the model as

$$Y^* = X^* b + e^*. \qquad (7.14)$$

Let Ω^* be the column space of X^*. This is our subspace of interest.

As mentioned above, usually in Model (7.14) there is no longer an intercept parameter; i.e., we have a case of **regression through the origin**. This is an explicit assumption for the following discussion; that is, we assume that the column space of X^* does not contain a column of ones. Because the rank-based estimators minimize a pseudo-norm we have

$$\|\boldsymbol{Y}^* - \boldsymbol{X}^*\boldsymbol{\beta}\|_\varphi \quad = \quad \sum_{i=1}^n a(R(y_i^* - \boldsymbol{x}_i^{*\prime}\boldsymbol{\beta}))(y_i^* - \boldsymbol{x}_i^{*\prime}\boldsymbol{\beta}) \tag{7.15}$$

$$= \quad \sum_{i=1}^n a(R(y_i^* - (\boldsymbol{x}_i^* - \overline{\boldsymbol{x}}^*)'\boldsymbol{\beta}))(y_i^* - (\boldsymbol{x}_i^* - \overline{\boldsymbol{x}}^*)'\boldsymbol{\beta})$$

$$= \quad \sum_{i=1}^n a(R(y_i^* - \alpha - (\boldsymbol{x}_i^* - \overline{\boldsymbol{x}}^*)'\boldsymbol{\beta}))(y_i^* - \alpha - (\boldsymbol{x}_i^* - \overline{\boldsymbol{x}}^*)'\boldsymbol{\beta}),$$

where $\boldsymbol{x}_i^*$ is the ith row of $\boldsymbol{X}^*$ and $\overline{\boldsymbol{x}}^*$ is the vector of column averages of $\boldsymbol{X}^*$. Hence, based on this result, the estimator of the regression coefficients based on the R fit of Model (7.14) is estimating the regression coefficients of the centered model, i.e., the model with the design matrix $\boldsymbol{X}_c = \boldsymbol{X}^* - \overline{\boldsymbol{X}^*}$. Thus the rank-based fit of the vector of regression coefficients is in the column space of $\boldsymbol{X}_c$, not the column space of $\boldsymbol{X}^*$.

Dixon and McKean (1996) proposed the following[3] solution. Assume that (7.14) is the true model, but first obtain the rank-based fit of the model:

$$\boldsymbol{Y}^* = \boldsymbol{1}\alpha_1 + \boldsymbol{X}^*\boldsymbol{b} + \boldsymbol{e}^* = [\boldsymbol{1} \ \boldsymbol{X}^*] \begin{bmatrix} \alpha_1 \\ \boldsymbol{b} \end{bmatrix} + \boldsymbol{e}^*, \tag{7.16}$$

where the true α_1 is 0. Let $\boldsymbol{U}_1 = [\boldsymbol{1} \ \boldsymbol{X}^*]$ and let Ω_{U_1} denote the column space of $\boldsymbol{U}_1$. Let $\widehat{\boldsymbol{Y}}_{U_1} = \boldsymbol{1}\widehat{\alpha}_1 + \boldsymbol{X}^*\widehat{\boldsymbol{b}}$ denote the R fitted value based on the fit of Model (7.16), where, as usual, the intercept parameter α_1 is estimated by the median of the residuals of the rank-based fit. Note that the subspace of interest Ω^* is a subspace of Ω_1, i.e., $\Omega^* \subset \Omega_{U_1}$. Secondly, project this fitted value onto the desired space Ω^*; i.e., let $\widehat{\boldsymbol{Y}}^* = \boldsymbol{H}_{\Omega^*}\widehat{\boldsymbol{Y}}_{U_1}$, where $\boldsymbol{H}_{\Omega^*} = \boldsymbol{X}^*(\boldsymbol{X}^{*\prime}\boldsymbol{X}^*)^{-1}\boldsymbol{X}^{*\prime}$. Thirdly, and finally, estimate $\boldsymbol{b}$ by solving the equation

$$\boldsymbol{X}^*\widehat{\boldsymbol{b}} = \widehat{\boldsymbol{Y}}^*, \tag{7.17}$$

that is, $\widehat{\boldsymbol{b}}^* = (\boldsymbol{X}^{*\prime}\boldsymbol{X}^*)^{-1}\boldsymbol{X}^{*\prime}\widehat{\boldsymbol{Y}}_{U_1}$. This is the rank-based estimate for Model (7.17).

The asymptotic variance of $\widehat{\boldsymbol{b}}^*$ is given by

$$\begin{aligned} \mathrm{AsyVar}(\widehat{\boldsymbol{b}}^*) \quad = \quad & \tau_S^2(\boldsymbol{X}^{*\prime}\boldsymbol{X}^*)^{-1}\boldsymbol{X}^{*\prime}\boldsymbol{H}_1\boldsymbol{X}^*(\boldsymbol{X}^{*\prime}\boldsymbol{X}^*)^{-1} \\ & + \tau_\varphi^2(\boldsymbol{X}^{*\prime}\boldsymbol{X}^*)^{-1}\boldsymbol{X}^{*\prime}\boldsymbol{H}_{\boldsymbol{X}_c}\boldsymbol{X}^*(\boldsymbol{X}^{*\prime}\boldsymbol{X}^*)^{-1}, \end{aligned} \tag{7.18}$$

where $\boldsymbol{H}_1$ and $\boldsymbol{H}_{\boldsymbol{X}_c}$ are the projection matrices onto a column of ones and the column space of $\boldsymbol{X}_c$, respectively.

We have written an R function which obtains the weighted rank-based fit. The function has the following arguments

---

[3]See, also, Page 287 of Hettmansperger and McKean (2011).

```
> args(wtedrb)
```

```
function (x, y, wts = diag(rep(1, length(y))), scores = wscores)
NULL
```

where y and x are the R vector of responses and design matrix, respectively. These responses and design matrix are the non-weighted values as in Model (7.13). The weights are in the matrix wts with the default as the identity matrix. The scores option allows for different scores to be used with the Wilcoxon scores as the default. Values returned are: fitted values ($yhatst), residuals ($ehatst), estimate of regression coefficients ($bstar), standard errors of estimates ($se), and estimated covariance matrix of estimates ($vc).

The two main applications for the function wtedrb are:

1. Models of the form (7.13) for which a weighted regression is desired. Let W denote the weight matrix. Then the model to be fitted is given in expression (7.14). In this case the matrix X^* is the design matrix argument for the function wtedrb and the matrix W is the weight matrix. It is assumed that a vector of ones is not in the column space of X^*.

2. A **regression through the origin** is desired. In this case, the model is

$$Y = X\beta + e = U_1 b + e, \tag{7.19}$$

   where the column space of X does not contain a vector of ones while that of U_1 does. For this model, the design matrix argument is X and there is no weight matrix; i.e., the default identity matrix is used. Some cautionary notes on using this model are discussed in Exercise 7.9.9.

Exercise 7.9.10 discusses a weighted regression. The next example serves as an illustration of regression through the origin.

Example 7.4.1 (Crystal Data). Hettmansperger and McKean (2011) discuss a dataset[4] where regression through the origin was deemed important. The response variable y is the weight of a crystalline form of a certain chemical compound and the independent variable x is the length of time that the crystal was allowed to grow. For convenience, we display the data in Table 7.2.

The following code segment computes the rank-based fit of this model (the responses are in the vector y and the vector for the independent variable is in x). Note that x is the designed matrix used and that there is no weight argument (the default identity matrix is used). Figure 7.10 contains the scatterplot of the data overlaid by the rank-based fit.

```
> wtedfit <- wtedrb(x,y)
> wtedfit$bstar
```

---

[4]See Graybill and Iyer (1994) for initial reference for this dataset.

TABLE 7.2

Crystal Data.

Time (hours)	2	4	6	8	10	12	14
Weight (grams)	0.08	1.12	4.43	4.98	4.92	7.18	5.57
Time (hours)	16	18	20	22	24	26	28
Weight (grams)	8.40	8.881	10.81	11.16	10.12	13.12	15.04

```
        [,1]
[1,] 0.5065172
```

```
> wtedfit$se
```

```
[1] 0.02977037
```

The rank-based estimate of slope is 0.507 with a standard error of 0.030; hence, the result is significantly different from 0. ∎

Remark 7.4.1. Weighted rank-based estimates can be used to fit general

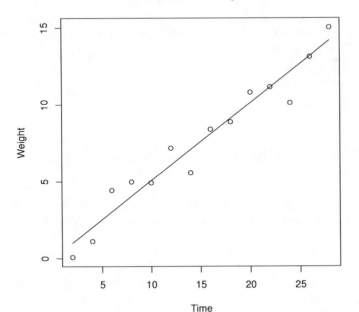

Rank–Based Fit of Crystal Data

FIGURE 7.10

Scatterplot of the Crystal data overlaid by the rank-based fit.

linear models where the variance of the random errors is an $n \times n$ matrix $\boldsymbol{\Sigma} > 0$. In this case the weight matrix is $\boldsymbol{\Sigma}^{-1/2}$, where $\boldsymbol{\Sigma}^{-1/2} = \boldsymbol{\Gamma}\boldsymbol{\Lambda}^{-1/2}\boldsymbol{\Gamma}'$ and $\boldsymbol{\Lambda}$ and $\boldsymbol{\Gamma}$ are respectively the diagonal matrix of eigenvalues and eigenvectors of $\boldsymbol{\Sigma}$. Generally in practice, $\boldsymbol{\Sigma}$ has to be estimated; see Bilgic (2012) for such an iterated reweighted procedure for hierarchical linear models. Dixon and McKean (1996) and Carroll and Ruppert (1982) developed robust (R and M, respectively) for iteratively estimating the regression coefficients and the weights when $\boldsymbol{\Sigma}$ is diagonal and the form of the heteroscedasticity is known; for example, when scale varies directly with response. ∎

7.5   Linear Models with Skew Normal Errors

The family of skew normal distributions consists of left and right skewed distributions along with the normal distribution. The pdfs in this family are of the form

$$f(x; \alpha) = 2\phi(x)\Phi(\alpha x), \tag{7.20}$$

where the parameter α satisfies $-\infty < \alpha < \infty$ and $\phi(x)$ and $\Phi(x)$ are the pdf and cdf of a standard normal distribution, respectively. If a random variable X has this pdf, we say that X has a standard skew normal distribution with parameter α and write $X \sim SN(\alpha)$. If $\alpha = 0$, then X has a standard normal distribution. Further X is distributed left skewed if $\alpha < 0$ and right skewed if $\alpha > 0$. This family of distributions was introduced by Azzalini (1985), who discussed many of its properties. We are interested in using this family of distributions for error distributions in the linear model,

$$Y_i = f(\boldsymbol{x}_i; \boldsymbol{\beta}) + e_i, \quad i = 1, 2, \ldots, n, \tag{7.21}$$

where $e = b\epsilon_i$, where $\epsilon_i \sim SN(\alpha)$, for some $-\infty < \alpha < \infty$, and the scale parameter $b > 0$. We next discuss rank-based fits and inference for such models; see McKean and Kloke (2014) for more information.

Since these fits are scale equivariant, there is no need to estimate the scale parameter b. Likewise, for inference on the vector of parameters $\boldsymbol{\beta}$ there is no need to estimate the shape parameter α. Wilcoxon scores could be used or the bent scores designed for skewed error distributions. However, to get an idea of what scores to consider, we next discuss the optimal scores for a specified α.

To obtain the optimal rank-based scores, because of equivariance, we need only the form (down to scale and location) of the pdf. So for the derivation of the scores, assume that the random variable $X \sim SN(\alpha)$ with pdf (7.20). Then as Exercise 7.9.12 shows:

$$-\frac{f'(x; \alpha)}{f(x; \alpha)} = x - \frac{\alpha\phi(\alpha x)}{\Phi(\alpha x)}. \tag{7.22}$$

Denote the inverse of the cdf of X by $F^{-1}(u; \alpha)$. Then the optimal score function for X is

$$\varphi_\alpha(u) = F^{-1}(u; \alpha) - \frac{\alpha\phi(\alpha F^{-1}(u; \alpha))}{\Phi(\alpha F^{-1}(u; \alpha))}. \tag{7.23}$$

For all values of α, this score function is strictly increasing over the interval $(0, 1)$; see Azzalini (1985). As expected, for $\alpha = 0$, the scores are the normal scores introduced in Chapter 3. Due to the first term on the right-side of expression (7.23), all the score functions in this family are unbounded, indicating that the skew normal family of distributions is light-tailed. Thus the influence functions of the rank-based estimators based on scores in this family are unbounded in the Y-space and, hence, are not robust. Recall in our discussion of the normal scores ($\alpha = 0$), which are optimal for normal pdfs, that the corresponding rank-based estimator is technically robust. The sensitivity analysis of Section 7.5.1 confirms this technical robustness for the skew normal scores, also.

We also need the derivative of (7.23) to complete the installation of these scores in `scores`. Let $l(x)$ denote the function defined in expression (7.22). Then as Exercise 7.9.14 shows, the derivative of the optimal score function is

$$\varphi'_\alpha(u) = l'[F^{-1}(u; \alpha)]\frac{1}{2\phi[F^{-1}(u; \alpha)]\Phi[\alpha F^{-1}(u; \alpha)]}, \tag{7.24}$$

where

$$l'(x) = 1 + \frac{\alpha^2\phi(\alpha x)[\alpha x\Phi[\alpha F^{-1}(u; \alpha)] + \phi(\alpha x)]}{\Phi^2(\alpha x)}.$$

Hence, to install this class of scores in `Rfit`, we only need the computation of the quantiles $F^{-1}(u; \alpha)$. Azzalini (2014) developed the R package `sn` which computes the quantile function $F^{-1}(u; \alpha)$ and, also, the corresponding pdf and cdf. The command `qsn(u,alpha=alpha)` returns $F^{-1}(u; \alpha)$, for $0 < u < 1$. We have added the class `skewns` to the book package `npsm`. Figure 7.11 displays these skew normal scores for $\alpha = -7, 1$, and 5 in the right panels of the figure and the corresponding pdf in the left panels.

Note that the pdf for $\alpha = -7$ is left skewed while those for positive α values are right skewed. Unsurprisingly, the pdf for $\alpha = 1$ is closer to symmetry than the pdfs of the others. The score function for the left-skewed pdf emphasizes relatively the right-tails over the left-tails, while the reverse is true for the right-skewed pdfs.

7.5.1 Sensitivity Analysis

For the sensitivity analysis, we generated $n = 50$ observations from a linear model of the form $y_i = x_i + e_i$, where x_i has a $N(0, 1)$ distribution and e_i has a $N(0, 10^2)$ distribution. The x_is and e_is are all independent. The generated data can be found in the data file `sensxy`. We added outliers of the form

$$y_{50} \leftarrow y_{50} + \Delta, \tag{7.25}$$

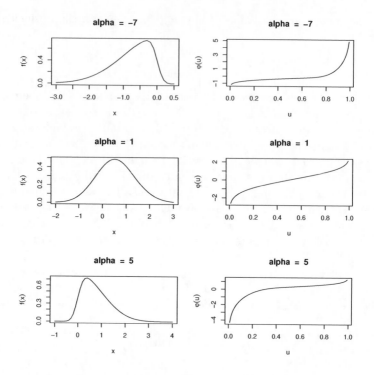

FIGURE 7.11
These plots display the pdfs of the three skew normal distributions with shape
parameter $\alpha = -7, 1$, and 5, along with the corresponding optimal scores.

where Δ ranges through the values in the top row of Table 7.3. For an estimator
$\hat{\beta}$, its sensitivity curve at Δ is

$$S(\Delta; \hat{\beta}) = \hat{\beta} - \hat{\beta}(\Delta), \qquad (7.26)$$

where $\hat{\beta}$ and $\hat{\beta}(\Delta)$ denote the estimates of β on the original and modified
data (7.25), respectively. We obtained sensitivity curves for the estimators:
Wilcoxon, normal scores, skew normal ($\alpha = 3$), skew normal ($\alpha = 5$), skew
normal ($\alpha = 7$), and maximum likelihood estimates (mle). The mles were
computed by the package sn. The following code segment illustrates setting
of the parameter in skewns:

```
> s5 <- skewns;  s5@param <- c(5)
```

That is, we first obtain a copy of the skew normal scores object (skewns) and
then we set the value of the α parameter to 5.

For all values of Δ, the changes in all of the the rank-based estimates were
less than 0.004. Thus the rank-based skew-normal estimators, including the

normal scores estimator, exhibited technical robustness for this study. On the other hand, the mle was sensitive to the values of Δ. We show these changes in Table 7.3; hence, for this study, the mle was not robust.

TABLE 7.3
Values of the Sensitivity Function for the mle.

Δ	0	20	40	60	80	100	1000	2000
mle	0.00	−0.07	−0.07	−0.00	0.12	0.30	−5.80	−6.32

7.5.2 Simulation Study

We conclude this section with the results of a small simulation study concerning rank-based procedures based on skew normal scores. The model simulated is

$$y_i = \beta_0 + \beta x_i + \theta c_i + e_i, \qquad (7.27)$$

where x_i is distributed $N(0,1)$; e_i is distributed from a selected error distribution; $i = 1, \ldots, 100$; the x_is and e_is are all independent; and the variable c_i is a treatment indicator with values of either 0 or 1. We selected two error distributions for the study. One is a skew normal distribution with shape parameter $\alpha = 5$ while the other is a contaminated version of a skew normal. The contaminated errors are of the form

$$e_i = (1 - I_{\epsilon,i})W_i + I_{\epsilon,i}V_i, \qquad (7.28)$$

where W_i has a skew normal distribution with shape parameter $\check{\alpha} = 5$, V_i has $N(\mu_c = 10, \sigma_c^2 = 36)$ distribution, $I_{\epsilon,i}$ has a binomial $(1, \epsilon = 0.15)$ distribution, and W_i, V_i, and $I_{\epsilon,i}$ are all independent. This contaminated distribution is skewed with heavy right-tails. The design is slightly unbalanced with $n_1 = 45$ and $n_2 = 55$. Without loss of generality β, θ, and β_0 were set to 0.

For procedures, we selected 7 rank-based procedures based on skew normal scores with the respective values of α set at $2, 3, \ldots, 8$; the Wilcoxon procedure; and the mle procedure. Hence, the asymptotically efficient rank-based procedure (score with $\alpha = 5$) is one of the selected procedures. The empirical results presented are the empirical AREs, which for each estimator is the ratio of the empirical mean-square error (MSE) of the mle to the empirical MSE of the estimator; hence, values of this ratio less than 1 are favorable to the mle while values greater than 1 are favorable to the estimator. Secondly, we present the empirical confidence coefficients for nominal 95% confidence intervals. For all the procedures, we chose asymptotic confidence intervals of the form $\hat{\beta} \pm 1.96SE(\hat{\beta})$. We used a simulation size of 10,000.

The results are presented in Table 7.4. For the skew normal errors, for both parameters β and θ, all the rank-based estimators except the Wilcoxon estimator are more efficient than the mle estimator. Note that the most efficient estimator for both β and θ is the rank-based estimator based on the

TABLE 7.4

Summary of Results of Simulation Study of Rank-Based Procedures
and the Maximum Likelihood Procedure for the Skew Normal
Distribution with Shape $\alpha = 5$ and a Skew Normal Contaminated
Distribution.

| | Skew Normal Errors | | | | Contaminated Errors | | | |
| | β | | θ | | β | | θ | |
Proced.	ARE	Conf.	ARE	Conf.	ARE	Conf.	ARE	Conf.
rb $\alpha = 2$	1.02	0.96	1.04	0.96	6.61	0.98	10.84	0.98
rb $\alpha = 3$	1.09	0.96	1.11	0.96	7.43	0.97	12.24	0.98
rb $\alpha = 4$	1.13	0.96	1.15	0.96	7.79	0.97	12.91	0.98
rb $\alpha = 5$	1.14	0.96	1.16	0.96	7.85	0.96	13.10	0.97
rb $\alpha = 6$	1.13	0.95	1.16	0.96	7.73	0.96	13.02	0.97
rb $\alpha = 7$	1.11	0.95	1.14	0.95	7.49	0.95	12.72	0.97
rb $\alpha = 8$	1.09	0.95	1.12	0.95	7.17	0.95	12.30	0.96
rb Wil.	0.78	0.95	0.79	0.95	4.70	0.96	7.56	0.97
mle	1.00	0.93	1.00	0.93	1.00	0.96	1.00	0.99

score function with $\alpha = 5$; although, the result is not significantly different than the results for a few of the nearby (α close to 5) rank-based estimators. In terms of validity, the empirical confidences of all the rank-based estimators are close to the nominal confidence of 0.95. In this study the mle is valid, also.

For the contaminated error distribution, the rank-based estimators are much more efficient than the mle procedure. Further, the estimator with scores based on $\alpha = 5$ is still the most empirically powerful in the study. It has empirical efficiency of 785% relative to the mle for β and 1310% for θ. All the rank-based procedures based on skew normal scores display technical robustness in this study. Even the Wilcoxon procedure is over 400% more efficient than the mle.

7.6   A Hogg-Type Adaptive Procedure

Score selection was discussed in Section 3.5.1 for the two-sample problem. The two-sample location model, though, is a linear model (see expression (3.23)); hence, the discussion on scores in Chapter 3 pertains to regression models of this chapter also. Thus, for Model (7.1) if we assume that the pdf of the distribution of the errors is $f(t) = f_0[(t - a)/b]$, where f_0 is known and a and b are not, then the scores generated by the score function

$$\varphi_{f_0}(u) = -\frac{f_0'(F_0^{-1}(u))}{f_0(F_0^{-1}(u))};\qquad(7.29)$$

lead to fully efficient rank-based estimates (asymptotically equivalent to maximum likelihood estimates). For example, if we assume that the random errors of Model (7.1) are normally distributed then rank-based fits based on normal scores are asymptotically equivalent to LS estimates. Other examples are discussed in Sections 7.5 and 3.5.1. Further, a family of score functions for log-linear models is discussed in Section 6.4.

Suppose, though, that we do not know the form. Based on the derivation of the optimal scores (7.29) given in Hettmansperger and McKean (2011), estimates based on scores "close" to the optimal scores tend to have high efficiency. A practical approach is to select a family of score functions which are optimal for a rich class of distributions and then use a data driven "selector" to choose a score from this family for which to obtain the rank-based fit of the linear model. We call such a procedure a data driven **adaptive scheme**. This is similar to Hogg's adaptive scheme for the two-sample problem (see Section 3.6), except our interest here is in obtaining a good fit of the linear model and not in tests concerned with location parameters.

Adaptive schemes should be designed for the problem at hand. For example, perhaps it is clear that only right skewed distributions for the random errors need to be considered. In this case, the family of scores should include scores appropriate for right skewed distributions.

For discussion, we consider a generic Hogg-type adaptive scheme designed for light to heavy tailed distributions which can be symmetric or skewed (both left and right). The scheme utilizes the class of bent (Winsorized Wilcoxons) scores discussed in detail in Section 3.5.1. As discussed in Chapter 3, the four types of bent scores are appropriate for our family of distributions of interest. For example, consider the `bentscores4`. These are optimal for distribution with a "logistic" middle and "exponential" tails. Scores corresponding to heavier tailed distributions have larger intervals where the score function is flat. Such scores are optimal for symmetric distributions if the bends are at c and $1 - c$, for $0 < c < (1/2)$; else, they are optimal for skewed distributions. If the distribution has a longer right than left tail then correspondingly the optimal score will have a longer flat interval on the right than on the left.

For our scheme, we have selected the nine bent scores which are depicted in Figure 7.12. The scores in the first column are for left skewed distributions, those in the second column are for symmetric distributions, while those in the third column are for right skewed distributions. The scores in the first row are for heavy tailed distributions, those in the second row are for moderate tailed distributions, and those in the third row are for light tailed distributions.

Recall that our goal is to fit a linear model; hence, the section of the score must be based on the residuals from an initial fit. For the initial fit, we have have chosen to use the Wilcoxon fit. Wilcoxon scores are optimal for the logistic distribution which is symmetric and of moderate tail weight, slightly heavier tails than those of a normal distribution. Let $\hat{e} = (\hat{e}_1, \ldots, \hat{e}_n)^T$, denote the vector of Wilcoxon residuals. As a selector, we have chosen the pair of statistics (Q_1, Q_2) proposed by Hogg. These are defined in expression (3.47)

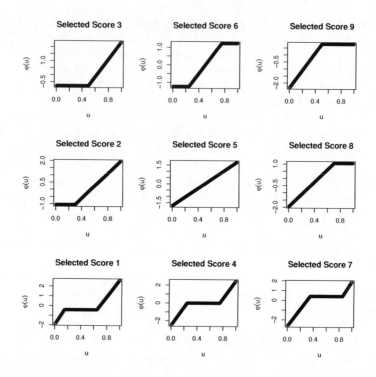

FIGURE 7.12
The nine bent scores for the generic Hogg-type adaptive scheme.

of Chapter 3. For the reader's convenience, we restate them here:

$$Q_1 = \frac{\bar{U}_{0.05} - \bar{M}_{0.5}}{\bar{M}_{0.5} - \bar{L}_{0.05}} \text{ and } Q_2 = \frac{\bar{U}_{0.05} - \bar{L}_{0.05}}{\bar{U}_{0.5} - \bar{L}_{0.5}}, \tag{7.30}$$

where $U_{0.05}$ is the mean of the Upper 5%, $M_{0.5}$ is the mean of the Middle 50%, and $L_{0.05}$ is the mean of the Lower 5% of the residuals $\hat{e}$. Recall that $Q1$ is a measure of skewness while $Q2$ is a measure of tail heaviness.

Cutoff values for the selection are required. In an investigation of this adaptive scheme for linear models, based on large simulation studies, Shomrani (2003) developed the following cutoff values:

$$
\begin{aligned}
c_{lq1} &= 0.36 + (0.68/n) \\
c_{uq1} &= 2.73 - (3.72/n) \\
c_{lq2} &= \begin{cases} 2.17 - (3.01/n) & n < 25 \\ 2.24 - (4.68/n) & n \geq 25 \end{cases} \\
c_{uq2} &= \begin{cases} 2.63 - (3.94/n) & n < 25 \\ 2.95 - (9.37/n) & n \geq 25 \end{cases}
\end{aligned}
$$

Note that the scores are numbered 1 through 9 in Figure 7.12. Using these numbers, the scheme's selected score is:

$$Q_1 \leq c_{lq1}, Q_2 \leq c_{lq2} \quad \text{Select} \quad \text{Score \#1}$$
$$Q_1 \leq c_{lq1}, c_{lq2} < Q_2 \leq c_{uq2} \quad \text{Select} \quad \text{Score \#2}$$
$$Q_1 \leq c_{lq1}, Q_2 > c_{uq2} \quad \text{Select} \quad \text{Score \#3}$$
$$c_{lq1} < Q_1 \leq c_{uq1}, Q_2 \leq c_{lq2} \quad \text{Select} \quad \text{Score \#4}$$
$$c_{lq1} < Q_1 \leq c_{uq1}, c_{lq2} < Q_2 \leq c_{uq2} \quad \text{Select} \quad \text{Score \#5}$$
$$c_{lq1} < Q_1 \leq c_{uq1}, Q_2 > c_{uq2} \quad \text{Select} \quad \text{Score \#6}$$
$$Q_1 > c_{uq1}, Q_2 \leq c_{lq2} \quad \text{Select} \quad \text{Score \#7}$$
$$Q_1 > c_{uq1}, c_{lq2} < Q_2 \leq c_{uq2} \quad \text{Select} \quad \text{Score \#8}$$
$$Q_1 > c_{uq1}, Q_2 > c_{uq2} \quad \text{Select} \quad \text{Score \#9} \tag{7.31}$$
$$\tag{7.32}$$

We have written an auxiliary R function `adaptor` which computes this adaptive scheme. The response vector Y and design matrix X form the input while the output includes both the initial (Wilcoxon) and selected fits, the scores selected and, for convenience, the number of the selected score. We illustrate its computation with the following examples.

Example 7.6.1 (Adaptive Scheme on Generated Exponential Errors). The first example consists of simulated data. We consider a regression model with two predictors each having a $N(0, 1)$ distribution and with sample size $n = 40$. For an error distribution, we chose an exponential distribution. We set all regression coefficients to 0, so that the model generated is:

$$y_i = 0 + 0 \cdot x_{i1} + 0 \cdot x_{i2} + e_i.$$

The following code segment computes the adaptive scheme for this dataset. It shows the respective summaries of the selected score fit and the Wilcoxon fit. The data are in the dataset `adapteg`.

```
> adapt <- adaptor(xmat,y)
> summary(adapt$fitsc)

Call:
rfit.default(formula = y ~ xmat, scores = sc, delta = delta,
    hparm = hparm)

Coefficients:
            Estimate Std. Error t.value  p.value
(Intercept)  0.607964  0.220940  2.7517 0.009127 **
xmat1       -0.082683  0.095334 -0.8673 0.391372
xmat2        0.028687  0.098423  0.2915 0.772319
---
```

Signif. codes: 0 '***' 0.001 '**' 0.01 '*' 0.05 '.' 0.1 ' ' 1

Multiple R-squared (Robust): 0.02974727
Reduction in Dispersion Test: 0.5672 p-value: 0.57197

```
> adapt$iscore

[1] 9

> summary(adapt$fitwil)
```

Call:
rfit.default(formula = y ~ xmat, delta = delta, hparm = hparm)

Coefficients:
```
            Estimate Std. Error t.value  p.value
(Intercept)  0.63612    0.22885  2.7796 0.008503 **
xmat1       -0.13016    0.14136 -0.9208 0.363142
xmat2        0.10842    0.14594  0.7429 0.462239
---
```
Signif. codes: 0 '***' 0.001 '**' 0.01 '*' 0.05 '.' 0.1 ' ' 1

Multiple R-squared (Robust): 0.04619702
Reduction in Dispersion Test: 0.89604 p-value: 0.41686

```
> precision <- (adapt$fitsc$tauhat/adapt$fitwil$tauhat)^2
> precision

[1] 0.4547999
```

In this case the adaptive scheme correctly chose score function #9; i.e., it selected the score for right-skewed error distributions with heavy tails. The ratio of the squared $\hat{\tau}$'s (the selected score function to the Wilcoxon score function) is 0.456; hence the selected fit is more precise in terms of standard errors than the Wilcoxon fit. ∎

For our second example we chose a real dataset.

Example 7.6.2 (Free Fatty Acid Data). In this dataset (ffa), the response is the free fatty acid level of 41 boys while the predictors are age (in months), weight (lbs), and skin fold thickness. It was initially discussed on page 64 of Morrison (1983) and more recently in Kloke and McKean (2012). The Wilcoxon Studentized residual and $q-q$ plots are shown in the top panels of Figure 7.13. Note that the residual plot indicates right skewness which is definitely confirmed by the $q-q$ plot. For this dataset, our adaptive scheme selected score function #8, bentscores2, with bend at $c = 0.75$, (moderately heavy tailed and right-skewed), which confirms the residual plot. The bottom panels of Figure 7.13 display the Studentized residual and q–q plots based on the fit from the selected scores.

For the next code segment the R matrix `xmat` contains the three predictors while the R vector `ffalev` contains the response. The summaries of both the initial Wilcoxon fit and the selected score fit are displayed along with the value of the precision.

```
> adapt <- adaptor(ffalev,xmat)
> summary(adapt$fitwil)

Call:
rfit.default(formula = y ~ xmat, delta = delta, hparm = hparm)

Coefficients:
               Estimate Std. Error t.value   p.value
(Intercept)   1.4900402  0.2692512  5.5340 2.686e-06 ***
xmatage      -0.0011242  0.0026348 -0.4267 0.6720922
xmatweight   -0.0153565  0.0038463 -3.9925 0.0002981 ***
```

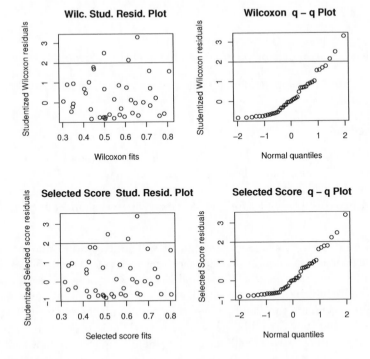

FIGURE 7.13
Wilcoxon (top pair) and the selected score (bottom pair) Studentized residual plots for Example 7.6.2.

```
xmatskinfold  0.2749014  0.1342149  2.0482 0.0476841 *
---
Signif. codes:  0 '***' 0.001 '**' 0.01 '*' 0.05 '.' 0.1 ' ' 1

Multiple R-squared (Robust): 0.3757965
Reduction in Dispersion Test: 7.42518 p-value: 0.00052

> summary(adapt$fitsc)

Call:
rfit.default(formula = y ~ xmat, scores = sc, delta = delta,
    hparm = hparm)

Coefficients:
               Estimate  Std. Error t.value   p.value
(Intercept)    1.41842877  0.23879002  5.9401 7.569e-07 ***
xmatage       -0.00078542  0.00232485 -0.3378  0.737394
xmatweight    -0.01538936  0.00339386 -4.5345 5.882e-05 ***
xmatskinfold   0.32728635  0.11842685  2.7636  0.008856 **
---
Signif. codes:  0 '***' 0.001 '**' 0.01 '*' 0.05 '.' 0.1 ' ' 1

Multiple R-squared (Robust): 0.397609
Reduction in Dispersion Test: 8.14063 p-value: 0.00027

> precision <- (adapt$fitsc$tauhat/adapt$fitwil$tauhat)^2
> precision

[1] 0.778572
```

Note that the precision estimate is 0.779, so the analysis based on the selected score is more precise than the initial Wilcoxon analysis. For example, the regression coefficient for the predictor skin fold thickness is highly significant for the bent score fit; while, for the Wilcoxon fit it is significant at the 5% level. ∎

Remark 7.6.1. Kapenga and McKean (1989) and Naranjo and McKean (1997) developed estimators for the score function $\varphi(u)$ based on residuals from a fit of the linear model. These and several other adaptive schemes including the Hogg-type scheme discussed above were compared in a large Monte Carlo study by Shomrani (2003). These schemes are ultimately for fitting the linear model and they are all based on residuals. So not surprisingly, their associated inference is somewhat liberal. In Al-Shomrani's study, however, the Hogg-type scheme was less liberal than the other schemes in the study. In general the Hogg-type scheme outperformed the others in terms of validity and empirical power. Okyere (2011) extended the Hogg-type scheme to mixed linear models. ∎

7.7 Nonlinear

In this section we present example code to obtain Wilcoxon estimates for the general nonlinear regression problem. The model that we consider is

$$y_i = f(\boldsymbol{\theta}; \boldsymbol{x}_i) + e_i \tag{7.33}$$

where $\boldsymbol{\theta}$ is a $k \times 1$ vector of unknown parameters, y_i is a response variable, and $\boldsymbol{x}_i$ is a $p \times 1$ vector of explanatory variables.

As a simple working example consider the simple nonlinear model

$$f(\theta; x_i) = \exp\{\theta x_i\}, \quad i = 1, \dots, n. \tag{7.34}$$

Suppose the errors are normally distributed. Setting $n = 25$, $\theta = 0.5$, and using uniformly distributed x's, a simulated example of this model is generated by the next code segment. Figure 7.14 displays a scatterplot of the data.

```
> n<-25
> theta<-0.5
> x<-runif(n,1,5)
> f<-function(x,theta) { exp(theta*x) }
> y<-f(x,theta)+rnorm(n,sd=0.5)
```

The rank-based fit of Model (7.33) is based on minimizing the same norm that was used for linear models. That is, for a specified score function $\varphi(u)$, the rank based estimate of $\boldsymbol{\theta}$ is

$$\hat{\boldsymbol{\theta}}_\varphi = \text{Argmin} \|\boldsymbol{Y} - f(\boldsymbol{\theta}; \boldsymbol{x})\|_\varphi, \tag{7.35}$$

where $\boldsymbol{Y}$ is the $n \times 1$ vector and the components of $f(\boldsymbol{\theta}; \boldsymbol{x})$ are the $f(\boldsymbol{\theta}; \boldsymbol{x}_i)$s. For the traditional LS estimate, the squared-Euclidean norm is used instead of $\|\cdot\|_\varphi$. The properties of the rank-based nonlinear estimator were developed by Abebe and McKean (2007). More discussion of these rank-based estimates can be found in Section 3.14 of Hettmansperger and McKean (2011), including the estimator's influence function. Based on this influence function, the rank-based estimator is robust in the $\boldsymbol{Y}$-space but not in the $\boldsymbol{x}$-space. The nonlinear HBR estimator developed by Abebe and McKean (2013) is robust with bounded influence in both the $\boldsymbol{Y}$-space and the $\boldsymbol{x}$-spaces. As in the case of linear models, the HBR estimator minimizes the weighted Wilcoxon norm; we recommend the high breakdown weights given in expression (7.9).

In this section, we discuss the Wilcoxon fit of a nonlinear model. The discussion for the HBR fit is in Section 7.7.4. We begin by discussing a simple Newton algorithm for the estimator and its subsequent computation using Rfit. For nonlinear fitting, the usual computational algorithm is a Gauss–Newton type procedure, which is based on a Taylor series expansion of $f(\boldsymbol{\theta}; \boldsymbol{x})$.

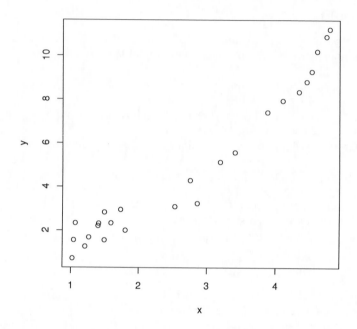

FIGURE 7.14
Scatterplot of the simulated data based on Model 7.34.

Let $\hat{\boldsymbol{\theta}}^{(0)}$ be an initial estimate of $\boldsymbol{\theta}$. The Jacobian at $\boldsymbol{\theta}$ is the $n \times k$ matrix

$$\boldsymbol{J}(\boldsymbol{\theta}) = \left[\frac{\partial f_i(\boldsymbol{\theta})}{\partial \theta_j}\right]. \tag{7.36}$$

The expansion of $f(\boldsymbol{\theta})$ about $\hat{\boldsymbol{\theta}}^{(0)}$ is

$$f(\boldsymbol{\theta}) \approx f(\hat{\boldsymbol{\theta}}^{(0)}) + J(\hat{\boldsymbol{\theta}}^{(0)})\boldsymbol{\Delta}, \tag{7.37}$$

where $\boldsymbol{\Delta} = \boldsymbol{\theta} - \hat{\boldsymbol{\theta}}^{(0)}$. Hence, an approximation to the norm is

$$\|\boldsymbol{Y} - f(\boldsymbol{\theta})\|_{\varphi} \approx \|\{[\boldsymbol{Y} - f(\hat{\boldsymbol{\theta}}^{(0)})] - J(\hat{\boldsymbol{\theta}}^{(0)})\boldsymbol{\Delta}(\boldsymbol{\theta})\}\|_{\varphi}. \tag{7.38}$$

Note that the quantity within the braces on the right side defines a linear model with the the quantity in brackets serving as the dependent variable, $J(\hat{\boldsymbol{\theta}}^{(0)})$ serving as the design matrix, and $\boldsymbol{\Delta}(\boldsymbol{\theta})$ serving as the vector of regression coefficients. For this linear model, let $\hat{\boldsymbol{\Delta}}$ be the rank-based estimate of $\boldsymbol{\Delta}$. Then the first step estimate of $\boldsymbol{\theta}$ is

$$\hat{\boldsymbol{\theta}}^{(1)} = \hat{\boldsymbol{\theta}}^{(0)} + \hat{\boldsymbol{\Delta}}. \tag{7.39}$$

Usually at this point of the algorithm, convergence is tested based on the relative increments in the estimates and the dispersion function. If convergence has not been achieved then $\hat{\boldsymbol{\theta}}^{(0)}$ is replaced by $\hat{\boldsymbol{\theta}}^{(1)}$ and the next step proceeds similar to the first step.

Often there is no intercept in the nonlinear model. In this case, the steps in the above algorithm consist of regressions through the origin. For rank-based estimation this is handled by the adjustment described in Section 7.4. Our R code for the ranked-based nonlinear fit, discussed in Section 7.7.2, automatically makes this adjustment. First, we discuss implementation of the rank-based nonlinear procedure.

7.7.1   Implementation of the Wilcoxon Nonlinear Fit

Consider the rank-based fit of the general nonlinear model (7.33) using Wilcoxon scores. Let $\hat{\boldsymbol{\theta}}_W$ denote the estimator (7.35). As shown in Abebe and McKean (2007), under general conditions, the asymptotic variance-covariance matrix of $\hat{\boldsymbol{\theta}}_W$ is given by

$$\tau_W^2 \left(J(\boldsymbol{\theta}) \right)^T J(\boldsymbol{\theta}) \right)^{-1}, \tag{7.40}$$

where $J(\boldsymbol{\theta})$ is the Jacobian evaluated at the true vector of parameters $\boldsymbol{\theta}$ and τ_W is the scale parameter given in expression (3.19). The only difference for the asymptotic variance of the LS estimator is that the variance of the random errors, σ^2, replaces τ_W^2. Hence, the asymptotic relative efficiency (ARE) of the Wilcoxon estimator relative to the LS estimator is σ^2/τ_W^2; i.e., the same ARE as in linear models. In particular, at normal errors this relative efficiency is 0.955.

Provided the Jacobian is a continuous function of $\boldsymbol{\theta}$, $J(\hat{\boldsymbol{\theta}}_W)^T J(\hat{\boldsymbol{\theta}}_W)$ is a consistent estimator of $J(\boldsymbol{\theta})^T J(\boldsymbol{\theta})$. Further, the same estimator of τ_W that we used in the linear model case (Koul et al. (1987)) but here based on the residuals $\hat{e}_W = \boldsymbol{Y} - f(\hat{\boldsymbol{\theta}}_W; \boldsymbol{x})$ is a consistent estimator of τ. Thus, the vector of standard errors of $\hat{\boldsymbol{\theta}}_W$ is

$$SE(\hat{\boldsymbol{\theta}}_W) = \text{Diagonal} \left\{ \left[\hat{\tau}_W^2 J(\hat{\boldsymbol{\theta}}_W)^T J(\hat{\boldsymbol{\theta}}_W)^{-1} \right]^{1/2} \right\}. \tag{7.41}$$

Note for future reference, these standard errors are essentially the standard errors of the approximate linear model on the last step of the Gauss–Newton algorithm.

7.7.2   R Computation of Rank-Based Nonlinear Fits

We have written R software for the computation of the nonlinear rank-based estimates which utilizes the Gauss–Newton algorithm described above. Currently, it has options for the Wilcoxon and HBR fits. The function is `wilnl` and is included in the package `rbnl`.[5] Its defining R statement with default values is

```
wilnl = function(x,y,theta0,fmodel,jmodel,numstp=50,
                 eps=.001,wts.type="WIL",
                  intest="HL",intercept=FALSE)
```

Definitions of these arguments are:

- `y` is the $n \times 1$ vector of responses and `x` is the $n \times p$ matrix of predictors.

- `theta0` is the initial estimate (starting value) of $\boldsymbol{\theta}$. The routine assumes it to be a $k \times 1$ **matrix**.

- `fmodel` and `jmodel` are user supplied R functions, one for the model and the other for the Jacobian. The arguments to these functions are matrices. These are described most easily by the ensuing discussion of examples.

- `numstp` and `eps` are the total number of Newton steps and the tolerance for stopping, respectively.

- `wts.type="WIL"` or `wts.type="HBR"` obtain, respectively, in this case, the Wilcoxon or the HBR rank-based nonlinear fits.

- The rank-based algorithm uses an estimate of a (pseudo) intercept. It is either the Hodges–Lehmann estimate, `intest="HL"` or the medium, `intest="MED"`. We recommend the default value `intest="HL"`, which generally leads to more efficient estimates, unless the data are highly skewed. In the later case we recommend the medium.

- Generally, nonlinear models do not have an intercept parameter. For such models, set the argument `int` at its default value, i.e., `int="NO"`. Occasionally, models do contain an intercept and, for these models, set `int` the value `int="YES"`. For models containing an intercept, one column of the Jacobian matrix consists of ones. In the user supplied Jacobian function, make this the first column of the Jacobian.

The returned `list` file includes the following items of interest: the estimate of $\boldsymbol{\theta}$, `$coef`; the standard errors of the estimates, `$se`; the estimate of the scale parameter τ, `$tauhat`; the residuals, `$resid`; and the results of each step, `$coll`, (the step values of $\boldsymbol{\theta}$ and $\|\boldsymbol{\theta}\|_2^2$).

The user supplied functions are most easily described by discussing a few examples. Consider first the simple working model (7.34). Recall that the nonlinear function is $f(\theta, x) = \exp\{\theta x\}$. The arguments are matrices `x` and `theta`. In the example `x` is 25×1 and `theta` is 1×1. The following model function, `expmod`, suffices:

```
expmod <- function(x,theta){ exp(x%*%theta) }
```

---

[5]See https://github.com/kloke/book for more information.

For the Jacobian there is only one partial derivative given by $\partial f / \partial \theta = x \exp\{\theta x\}$. Hence our Jacobian function, expjake, is

```
expjake <- function(x,theta){ x*exp(x%*%theta) }
```

The analysis of a simple dataset follows in the next example.

7.7.3 Examples

Example 7.7.1 (Computation of Rank-Based Estimates for Model (7.34)). The following rounded data are a generated realization of the Model (7.34).

x	4.4	4.3	1.4	2.2	3.1	4.2	2.0	3.1	4.0	1.9	2.9	1.4	3.2
y	8.4	8.4	1.8	3.2	4.3	7.6	3.3	4.8	7.5	2.4	4.0	1.9	6.0
x	2.5	4.4	3.6	4.3	1.5	1.0	4.1	1.1	1.9	4.0	1.1	2.2	
y	3.6	8.5	5.3	8.8	1.4	1.8	7.1	1.4	2.9	7.3	2.6	2.3	

For the following code segment, the 25×1 matrix x contains the x values while the vector y contains the y values. As a starting value, the true parameter $\theta = 0.5$ is used. For comparison, we computed the LS fit of this nonlinear model using the R function nls.

```
> expmod <- function(x,theta){ exp(x%*%theta) }
> expjake <- function(x,theta){ x*exp(x%*%theta) }
> fitwil <- wilnl(x,y,0.5,expmod,expjake)
> fitwil

Call:
wilnl(x = x, y = y, theta0 = 0.5, fmodel = expmod,
    jmodel = expjake)

Coefficients:
        [,1]
[1,] 0.49042

> summary(fitwil)

Call:
wilnl(x = x, y = y, theta0 = 0.5, fmodel = expmod,
    jmodel = expjake)

Coefficients:
        Estimate Std. Error t.value
[1,] 0.4904200  0.0050235  97.625

Number of iterations:  2
```

```
> fitls <- nls(y~exp(x*theta),start=list(theta=0.5))
> summary(fitls)
```

Formula: y ~ exp(x * theta)

Parameters:
```
       Estimate Std. Error t value Pr(>|t|)
theta 0.491728   0.004491   109.5   <2e-16 ***
---
```

Signif. codes: 0 '***' 0.001 '**' 0.01 '*' 0.05 '.' 0.1 ' ' 1

Residual standard error: 0.4636 on 24 degrees of freedom

Number of iterations to convergence: 2
Achieved convergence tolerance: 5.876e-06

The Wilcoxon and LS results are quite similar which is to be expected since the error distribution selected is normal. The Wilcoxon algorithm converged in three steps. ∎

For a second example, we altered this simple model to include an intercept.

Example 7.7.2 (Rank-Based Estimates for Model (7.34) with an Intercept). Consider the intercept version of Model (7.34); i.e., $f(\boldsymbol{\theta}; x) = \theta_1 + \exp\{\theta_2 x\}$. For this example we set $\theta_1 = 2$. We use the same data as in the last example, except that 2 is added to all the components of the vector y. The code segment follows. Note that both the model and Jacobian functions have been altered to include the intercept. As starting values, we chose the true parameters $\theta_1 = 2$ and $\theta_2 = 0.5$.

```
> y <- y + 2
> expmod <- function(x,theta){ theta[1]+exp(x%*%theta[2]) }
> expjake <- function(x,theta){ cbind(rep(1,length(x[,1])),
+              x*exp(x%*%theta[2])) }
> fitwil = wilnl(x,y,as.matrix(c(2,.5),ncol=1), expmod,
+              expjake,intercept=TRUE)
> fitwil

Call:
wilnl(x = x, y = y, theta0 = as.matrix(c(2, 0.5), ncol = 1),
    fmodel = expmod, jmodel = expjake, intercept = TRUE)

Coefficients:
          [,1]
[1,] 2.0436573
[2,] 0.4887782

> summary(fitwil)
```

```
Call:
wilnl(x = x, y = y, theta0 = as.matrix(c(2, 0.5), ncol = 1),
    fmodel = expmod, jmodel = expjake, intercept = TRUE)

Coefficients:
      Estimate Std. Error t.value
[1,] 2.0436573  0.1636433  12.489
[2,] 0.4887782  0.0080226  60.925

Number of iterations:  2
```

Note that the Wilcoxon estimate of the intercept is close to the true value of 2. For both parameters, the asymptotic 95% confidence intervals (± 1.96SE) trap the true values. ∎

The 4 Parameter Logistic Model

The 4 parameter logistic is a nonlinear model which is often used in pharmaceutical science for dose-response situations. The function is of the form

$$y = \frac{a - d}{1 + (x/c)^b} + d, \tag{7.42}$$

where x is the dose of the drug and y is the response. The exponent b is assumed to be negative; hence, as $x \to 0$ (0 concentration) $y \to d$ and as $x \to \infty$ (full concentration) $y \to a$. So a and d are the expected values of the response under minimum and maximum concentration of the drug, respectively. It follows that the value of $(a + d)/2$ is the 50% response rate and that this occurs at $x = c$. In terms of biological assays, the value c is called the IC_{50}, the amount of concentration of the drug required to inhibit a biological process by 50%. See Crimin et al. (2012) for discussion of this model and the robust Wilcoxon fit of it.

Usually in pharmaceutical science the dose of the drug is in log base 10 units. Also, this is an intercept model. Let $z = \log_{10} x$ and, to isolate the intercept, let $s = a - d$. Then the 4 parameter logistic model can be equivalently expressed as

$$Y_i = \frac{s}{1 + \exp\{b[z_i \log(10) - \log(c)]\}} + d + e_i, \quad i = 1, 2, \ldots, n. \tag{7.43}$$

In this notation, as $z \to -\infty$, $E(Y_i) \to d$ and as $z \to \infty$, $E(Y_i) \to a$. Figure 7.15 shows the LS and Wilcoxon fits of this model for a realization of the model discussed in Example 7.7.3. From this scatterplot of the data, guesstimates of the asymptotes a and d and the IC_{50} c are readily obtained for starting values. For the Jacobian, the four partial derivatives of the model function f are given by

$$\frac{\partial f}{\partial d} = 1$$

$$\frac{\partial f}{\partial s} = \frac{1}{1 + \exp\{b[z_i \log(10) - \log(c)]\}}$$

$$\frac{\partial f}{\partial c} = s\{1 + \exp\{b[z_i \log(10) - \log(c)]\}\}^{-2}$$
$$\times \left\{\frac{b}{c}\exp\{b[z\log(10) - \log(c)]\}\right\}$$

$$\frac{\partial f}{\partial b} = -s\{1 + \exp\{b[z_i \log(10) - \log(c)]\}\}^{-2}$$
$$\times \{[z\log(10) - \log(c)]\exp\{b[z\log(10) - \log(c)]\}\} \quad (7.44)$$

In the next example, we obtain the robust fit of a realization of Model 7.43.

Example 7.7.3 (The 4 Parameter Logistic). We generated a realization of size $n = 24$ from Model (7.43) with doses ranging from 0.039 to 80 with two repetitions at each dose. The data are in the set `eg4parm`. For this situation, normal random errors with standard deviation one were generated. The parameters were set at $a = 10$, $b = -1.2$, $c = 3$, and $d = 110$; hence, $s = -100$. The scatterplot of the data is displayed in Figure 7.15. The functions for the model and the Jacobian are displayed in the next code segment. This is followed by the computation of the LS and Wilcoxon fits. These fits are overlaid on the scatterplot of Figure 7.15. The segment of code results in a comparison of the LS and Wilcoxon estimates of the coefficients and their associated standard errors. Note that the parameter s was fit, so a transformation is needed to obtain the estimates and standard errors of the original parameters.

```
> func <- function(z,theta){
+       d = theta[1]; s = theta[2]; c = theta[3]; b = theta[4]
+       func <- (s/(1 + exp(b*(z*log(10) - log(c))))) + d
+       func
+ }
> jake = function(z,theta){
+       d = theta[1]; s = theta[2]; c = theta[3]; b = theta[4]
+       xp = 1 + exp(b*(z*log(10) - log(c)))
+       fd = 1; fs = 1/xp
+       fc = s*(xp^(-2))*((b/c)*exp(b*(z*log(10) - log(c))))
+       fb = -s*(xp^(-2))*((z*log(10) - log(c))*exp(b*(z*log(10) - log(c))))
+       jake = cbind(fd,fs,fc,fb); jake
+ }

> fitwil = wilnl(z,y, theta0,func,jake,intercept=TRUE)
> summary(fitwil)

Call:
wilnl(x = z, y = y, theta0 = theta0, fmodel = func,
    jmodel = jake, intercept = TRUE)
```

```
Coefficients:
     Estimate Std. Error t.value
fd 105.81755    5.60053 18.8942
   -95.07300   10.31646 -9.2157
     3.55830    0.83385  4.2673
    -1.81164    0.68261 -2.6540
```

Number of iterations: 7

```
> fitls <- nls(y~(s/(1 + exp(b*(z*log(10) - log(c))))) + d,
+ start=list( b =-1.2,  c = 3,d = 110,s = -100))
> summary(fitls)
```

Formula: y ~ (s/(1 + exp(b * (z * log(10) - log(c))))) + d

```
Parameters:
  Estimate Std. Error t value Pr(>|t|)
b  -2.1853     0.8663  -2.523 0.020230 *
c   3.0413     0.6296   4.831 0.000102 ***
d 105.4357     5.4575  19.319 2.09e-14 ***
s -93.0179     9.2404 -10.066 2.83e-09 ***
---
Signif. codes:   0 '***' 0.001 '**' 0.01 '*' 0.05 '.' 0.1 ' ' 1
```

Residual standard error: 16.02 on 20 degrees of freedom

Number of iterations to convergence: 11
Achieved convergence tolerance: 4.751e-06

```
> resid = fitwil$residuals
> yhat = fitwil$fitted.values
> ehatls <- summary(fitls)$resid
> yhatls <- y -ehatls
```

Note that the LS fit has been impacted by the outlier with the $\log_{10}$ dose at approximately 0.7. The LS function nls interchanged the order of the coefficients. The standard errors of the LS estimates are slightly less.

To demonstrate the robustness of the Wilcoxon fit, we changed the last response item from 11.33 to 70.0. The summary of the Wilcoxon and LS fits follows, while Figure 7.16 contains the scatterplot of the data and the overlaid fits.

```
> summary(fitwil)
```

```
Call:
wilnl(x = z, y = y, theta0 = theta0, fmodel = func,
    jmodel = jake, intercept = TRUE)
```

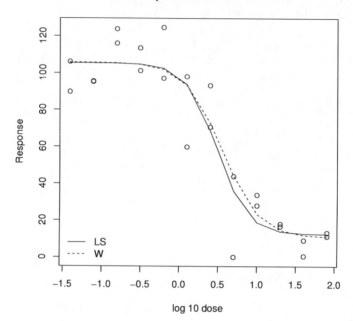

FIGURE 7.15
Scatterplot of 4 parameter logistic data overlaid by Wilcoxon (W) and least
squares (LS) nonlinear fits.

```
Coefficients:
       Estimate Std. Error t.value
fd  105.01857    6.08552 17.2571
    -89.58696   10.53863 -8.5008
      3.35834    0.81774  4.1068
     -2.20656    1.03166 -2.1388

Number of iterations:  3

> summary(fitls)

Formula: y ~ (s/(1 + exp(b * (z * log(10) - log(c))))) + d

Parameters:
   Estimate Std. Error t value Pr(>|t|)
b   -6.4036     9.2048  -0.696  0.49464
c    2.9331     0.7652   3.833  0.00104 **
```

```
d 101.9051      5.7583 17.697 1.10e-13 ***
s -79.1769      9.2046 -8.602 3.73e-08 ***
---
Signif. codes:  0 '***' 0.001 '**' 0.01 '*' 0.05 '.' 0.1 ' ' 1

Residual standard error: 19.8 on 20 degrees of freedom

Number of iterations to convergence: 32
Achieved convergence tolerance: 7.87e-06

>
```

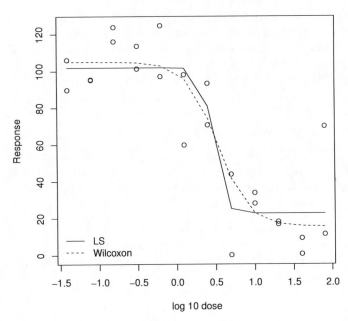

Least squares and Wilcoxon fits

FIGURE 7.16
Scatterplot of 4 parameter logistic changed data overlaid by Wilcoxon (W) and least squares (LS) nonlinear fits.

As Figure 7.16 shows, the LS fit was impaired by the outlier. The LS estimates of the parameters a and d changed by over a standard error and the LS estimate for b changed by 3 standard errors. On the other hand, the change in the Wilcoxon estimates was much less. ∎

7.7.4 High Breakdown Rank-Based Fits

As we mentioned earlier the Wilcoxon nonlinear estimator is much less sensitive to outliers in the response Y-space than the LS estimators, but, similar to LS, it is sensitive to outliers in the factor x-space. The weighted Wilcoxon HBR estimator, though, is not sensitive to such outliers. For Model (7.33), this estimator is given by

$$\hat{\boldsymbol{\theta}}_{HBR} = \text{Argmin} \|\boldsymbol{Y} - f(\boldsymbol{\theta}; \boldsymbol{x})\|_{HBR}, \qquad (7.45)$$

where $\|\cdot\|_{HBR}$ is the weighted norm given in expression (7.6). In our discussion, we use the high breakdown weights discussed in Section 7.2. Under general conditions, Abebe and McKean (2013) developed the asymptotic properties of the estimator $\hat{\boldsymbol{\theta}}_{HBR}$ including its asymptotic variance-covariance matrix. The Newton-type algorithm discussed for the rank-based estimators above works for the HBR estimator, also.

Computation of the HBR estimator is handled by the function `wilnl` using `wts="HBR"` as the argument for weights. As with the Wilcoxon nonlinear estimator, a consistent estimator of the variance-covariance matrix of the HBR estimator is obtained from the last linear step. The adjustment for the regression through the origin is handled similar to the Wilcoxon nonlinear estimator and, as with the Wilcoxon, the intercept estimator can be either the median of the residuals `int="MED"` or the Hodges–Lehmann `int="HL"`. We demonstrate the computation for the our simple working example.

Example 7.7.4 (HBR Estimator for Example 7.7.1). The following code segment uses the same data as in Example 7.7.1. In particular the 25×1 R matrix x contains the independent variable and the 25×1 R vector y contains the dependent variable.

```
> expmod <- function(x,theta){ exp(x%*%theta) }
> expjake <- function(x,theta){ x*exp(x%*%theta) }
> fithbr = wilnl(x,y,0.5,expmod,expjake,wts.type="HBR")
> fithbr$coef

          [,1]
[1,] 0.4904202

>
```

For this data there are no outliers in the x-space, so the HBR and Wilcoxon fits coincide. To demonstrate the robustness of the HBR estimator to outliers in factor space, we changed the last value of x from 2.3 to 8.0. The results are

```
> x[25,1] <- 8.0
> fitwil <- wilnl(x,y,0.5,expmod,expjake,wts.type="WIL")
> fitwil$coef
```

```
        [,1]
[1,] 0.2864679
```

```
> fithbr <- wilnl(x,y,0.5,expmod,expjake,wts.type="HBR")
> fithbr$coef
```

```
        [,1]
[1,] 0.488722
```

```
> fitls <- nls(y~exp(x*theta),start=list(theta=0.5))
> fitls$coef
```

NULL

Note that the Wilcoxon estimate of θ changed (absolutely) by 0.2, while the HBR estimate remained about the same. Also, the LS fit did not converge.
∎

The diagnostics, TDBETAS and CFITS, which differentiate among the LS, Wilcoxon, and HBR fits that were discussed for linear models in Section 7.3 extend straightforwardly to nonlinear models, including their benchmarks; see Abebe and McKean (2013) for details.

7.8   Time Series

Let $\{X_t\}$ be a sequence of random variables observed over time, $t = 1, 2, \ldots,$ n. A regression model frequently used in practice is the autoregressive model. This is a time series model where the observation at time t is a function of past observations plus some random noise. In this section, we discuss rank-based procedures for general order p autoregressive models. A related model consists of a linear model with random errors that follow a time series. Some discussion on robust procedures for these types of models is discussed in Section 6.6.3 of Hettmansperger and McKean (2011).

We say that X_t follows an autoregressive time series of order p, $X_t \sim$ AR(p), if

$$
\begin{aligned}
X_t &= \phi_0 + \phi_1 X_{t-1} + \phi_2 X_{t-2} + \cdots + \phi_p X_{t-p} + e_t \\
&= \phi_0 + \mathbf{Y}'_{t-1}\boldsymbol{\phi} + e_t, \quad t = 1, 2, \ldots, n
\end{aligned}
\tag{7.46}
$$

where $p \geq 1$, $\mathbf{Y}_{t-1} = (X_{t-1}, X_{t-2}, \ldots, X_{t-p})'$, $\boldsymbol{\phi} = (\phi_1, \phi_2, \ldots, \phi_p)'$, and $\mathbf{Y}_0$ is an observable random vector independent of $\mathbf{e}$. The stationarity assumption requires that the solutions to the following equation,

$$
x^p - \phi_1 x^{p-1} - \phi_2 x^{p-2} - \cdots - \phi_p = 0
\tag{7.47}
$$

lie in the interval $(-1, 1)$; see, for example, Box et al. (2008). We further assume that the components of e, e_t, are iid with cdf $F(x)$ and pdf $f(x)$, respectively.

Model (7.47) is a regression model with the tth response given by X_t and the tth row of the design matrix given by $(1, X_{t-1}, \ldots, X_{t-p})$, $t = p+1, \ldots, n$. Obviously the time series plot, X_t versus t, is an important first diagnostic. Also lag plots of the form X_t versus X_{t-j} are informative on the order p of the autoregressive series. We discuss determination of the order later.

As in Chapter 3, let $\varphi(u)$ denote a general score function which generates the score $a_\varphi(i) = \varphi[i/(n+1)]$, $i = 1, \ldots, n$. Then the rank-based estimate of ϕ is given by

$$
\begin{aligned}
\widehat{\phi} &= \operatorname{Argmin} D_\varphi(\phi) \\
&= \operatorname{Argmin} \sum_{t=p+1}^{n} a_\varphi[R(X_t - Y'_{t-1}\phi)](X_t - Y'_{t-1}\phi), \qquad (7.48)
\end{aligned}
$$

where $R(X_t - Y'_{t-1}\phi)$ denotes the rank of $X_t - Y'_{t-1}\phi$ among $X_1 - Y'_0\phi, \ldots, X_n - Y'_{n-1}\phi$. Koul and Saleh (1993) developed the asymptotic theory for these rank-based estimates. Outlying responses, though, also appear on the right side of the model; hence, error distributions with even moderately heavy tails produce outliers in factor space (points of high leverage). With this in mind, the HBR estimates of Section 7.2 should also be fitted. The asymptotic theory for the HBR estimators for Model (7.46) was developed by Terpstra et al. (2000) and Terpstra et al. (2001); see, also, Chapter 5 of Hettmansperger and McKean (2011) for discussion.

The rank-based and HBR fits of the AR(p) are computed by `Rfit`. As simulations studies have confirmed, the usual standard errors from regression serve as good approximations to the asymptotic standard errors. Also, Terpstra et al. (2003) developed Studentized residuals for rank-based fits based on an AR(1) model. They showed, though, that Studentized residuals from the rank-based fit of the regression model were very close approximations to the AR(1) Studentized residuals. In this section, we use the regression Studentized residuals. We have written a simple R function, `lagmat`, which returns the design matrix given the vector of responses and the order of the autoregressive series. We illustrate this discussion with a simple example of generated data.

Example 7.8.1 (Generated AR(2) Data). In this example we consider a dataset consisting of $n = 50$ variates generated form an AR(2) model with $\phi_1 = 0.6$, $\phi_2 = -0.3$, and random noise which followed a Laplace distribution with median 0 and scale parameter 10. The time series is plotted in the upper panel of Figure 7.17. The following R code segment obtains the Wilcoxon fit of the AR(2) model.

```
> data <- lagmat(ar2,2)
> x <- data[,1]
> lag1 <- data[,2]
```

```
> lag2 <- data[,3]
> wil <- rfit(x ~ lag1 + lag2)
> summary(wil)

Call:
rfit.default(formula = x ~ lag1 + lag2)

Coefficients:
            Estimate Std. Error t.value   p.value
(Intercept) -2.44173    2.97252 -0.8214  0.415731
lag1         0.70101    0.12115  5.7864 6.474e-07 ***
lag2        -0.36302    0.12280 -2.9563 0.004943 **
---
Signif. codes:  0 '***' 0.001 '**' 0.01 '*' 0.05 '.' 0.1 ' ' 1

Multiple R-squared (Robust): 0.399343
Reduction in Dispersion Test: 14.95899 p-value: 1e-05

> studresids <- rstudent(wil)
```

Note that the estimates of ϕ_1 and ϕ_2 are respectively 0.70 and -0.36 which are close to their true values of 0.6 and -0.3. With standard errors of 0.12, the corresponding 95% confidence intervals trap these true values. The lower panel of Figure 7.17 displays the plot of the Studentized Wilcoxon residuals. At a few time points the residual series is outside of the ± 2 bounds, but these are mild discrepancies. The largest in absolute value has the residual value of about -4 at time 8. We next compute the HBR fit of the AR(2) model and obtain the diagnostic TDBETAS between the Wilcoxon and HBR fits.

```
> hbr <- hbrfit(x~cbind(lag1,lag2))
> summary(hbr)

Call:
hbrfit(formula = x ~ cbind(lag1, lag2))

Coefficients:
                         Estimate Std. Error t.value   p.value
(Intercept)              -2.37583    2.92632 -0.8119  0.421133
cbind(lag1, lag2)lag1     0.70400    0.13865  5.0774 7.107e-06 ***
cbind(lag1, lag2)lag2    -0.36436    0.10811 -3.3704 0.001549 **
---
Signif. codes:  0 '***' 0.001 '**' 0.01 '*' 0.05 '.' 0.1 ' ' 1

Wald Test: 13.52353 p-value: 3e-05

> dnost <- fitdiag(cbind(lag1,lag2),x,est=c("WIL","HBR"))
> c(dnost$tdbeta,dnost$bmtd)
```

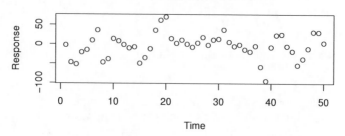

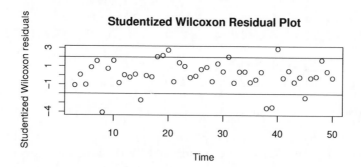

FIGURE 7.17
The upper panel shows the time series plot of the generated AR(2) data, while the lower panel is a plot of the Studentized Wilcoxon residuals from the fit of the AR(2) model.

[1] 0.001607918 0.750000000

Notice that the HBR fit is quite close to the Wilcoxon fit. This is confirmed by TDBETAS which has the value 0.002 (the benchmark is 0.75). ■

There are essentially two type of outliers for time series. The first type are the **innovative outliers** (IO). These can occur when the error distribution has heavy-tails. If an outlier occurs at time t ($|e_t|$ is large), then this generally leads to a response outlier, X_t, at time t, i.e., an outlier on the left-side of the model. In subsequent times it appears on the right-side and becomes incorporated into the model. These IO outliers generally lead to good points of high leverage. The dataset generated in Example 7.8.1 illustrates IO outliers. **Additive outliers** are a second type of outliers. These are patched into the series by a contaminating process, often leading to bad leverage points; see page 413 of Terpstra et al. (2001) for details. Both types of outliers occur in practice. For this reason, we recommend fitting both highly efficient and high breakdown rank-based estimates to the series and using the diagnostic TD-

BETAS to see if the fits differ. Terpstra et al. (2001) performed a large Monte Carlo study of rank-based and M estimates of time series involving both types of outliers. In all of the situations they simulated, the rank-based estimates either performed the best or nearly best over all procedures considered.

7.8.1 Order of the Autoregressive Series

In practice, before fitting the autoregressive time series model, (7.46), we must decide on its order p. One approach consists of subsequent fitting of models beginning with a model of large order. As discussed in Terpstra et al. (2001) for rank-based procedures, a more precise statement of the algorithm is:
First select a value of P of maximal order; i.e., the residual analysis shows that the model fits well. Next, select a level α for the testing. Then the algorithm is given by

(0) Set $p = P$.

(1) While $p > 0$, fit Model (7.46) with order p.

(2) Let $\phi_2 = (\phi_p, \ldots, \phi_P)'$. Then use the Wald test procedure to test $H_0 : \phi_2 = \mathbf{0}$ versus $H_A : \phi_2 \neq \mathbf{0}$.

(3) If H_0 is rejected then stop and declare p to be the order; otherwise, set $p = p - 1$ and go to (1).

We return to the last example and use this algorithm to determine the order of the AR model based on the HBR fit.

Example 7.8.2 (Generated AR(2) Data, Continued). We have written a R function, `arorder`, to implement the above algorithm. The sequential testing is based on rank-based Wald's test of $A\phi = \mathbf{0}$ where A is the constraint matrix as described in Step 2 of the algorithm. The user must select a maximum order of the AR model. For the example we have selected `maxp=4`. Besides this, as shown in the code segment, the estimates and the covariance matrix of the estimates are inputted. It returns the stepwise results of the algorithm. We illustrate its use for the HBR fit.

```
> data <- lagmat(ar2,4)
> x <- data[,1]
> xmat <- data[,2:(4+1)]
> hbr <- hbrfit(x~xmat)
> varcov = vcov(hbr,details=T)
> est <- hbr$coef
> alg <- arorder(length(x),4,est,varcov)
> alg$results

     [,1]      [,2]          [,3]
   4 0.8331220 0.366707515
```

```
3 0.4832491 0.620244664
2 4.6797785 0.006679725
```

In this case, the algorithm correctly identifies the order of the autoregressive series which is 2. ■

A rank-based adaption of Graybill's algorithm performed well in a simulation study by Terpstra and McKean (2005). A similar algorithm for the order of a polynomial regression model was discussed in Section 4.7.1.

7.9 Exercises

7.9.1. To see the effect on fits that "good" and "bad" points of high leverage can have, consider the following dataset:

x	1	2	3	4	5	6	7	8	9	10	20
y	5	7	6	14	14	25	29	33	31	41	75
y_2	5	7	6	14	14	25	29	33	31	41	20

The point $x = 20$ is a point of high leverage. The data for y (rounded) are realizations from the model $y = 4x + e$, where e has a $N(0, 9)$ distribution. Hence, the the value of y follows the model and is a "good" point of high leverage. Notice that y_2 is the same as y, except the last component of y_2 has been changed to 20 and, thus, is a "bad" point of high leverage.

    (a) Obtain the scatterplot for x and y, the Wilcoxon and HBR fits, and overlay these fits on the scatterplot.

    (b) Obtain the scatterplot for x and y_2, the Wilcoxon and HBR fits, and overlay these fits on the scatterplot.

    (c) Comment on the differences among the fits and plots.

7.9.2. Consider the datasets in Exercise 7.9.1.

    (a) Using the function `fitdiag`, obtain the diagnostics TDBETAS and CFITS for the set x and y. Plot CFITS versus Case. Comment.

    (b) Next obtain the diagnostics and plot for the set x and y_2. Comment.

7.9.3. There is some loss of efficiency when using the HBR fit instead of the Wilcoxon for "good" data. Verify this for a simulation of the model $y = 4x + e$, where e has a $N(0, 625)$ distribution and $x = 1 : 20$, using 10,000 simulations.

7.9.4. Using the set up Exercise 7.9.3, check the validity of the 95% confidence intervals for β_1 obtained by the Wilcoxon and HBR fits.

7.9.5. Obtain the Wilcoxon and HBR fits for the simple "good" set of data discussed at the beginning of Section 7.3. Then determine the diagnostics TDBETAS and CFITS. Is there a difference in fits? Next obtain the plot of robust distances and TDBETAS between the LTS fit and the Wilcoxon fit. Comment.

7.9.6. Hawkins et al. (1984) presented a simulated dataset consisting of 75 data points and three predictors. The first 14 points are outliers in the X-space while the remaining 61 points follow a linear model. Of the 14 outliers, the first 10 do not follow the model while the next 4 do follow the model. The dataset is in the `hawkins`.

(a) Obtain the Wilcoxon fit and plot its Studentized residuals versus Case number. Comment.

(b) Obtain the HBR fit and plot its Studentized residuals versus Case number. Comment.

(c) Obtain the diagnostics TDBETAS and CFITS between the HBR and Wilcoxon fits. Plot CFITS versus Case. Did the diagnostics discern the embedded outlier structure for this dataset?

7.9.7. Use the `bonds` data to examine the relationship between the bid prices for US treasury bonds (`BidPrice`) and the size of the bond's periodic payment rate (`CouponRate`). Consider `BidPrice` as the response variable and `CouponRate` as the explanatory variable. Obtain the Wilcoxon and HBR fits of this dataset. Using Studentized residuals and the diagnostics TDBETAS and CFITS, show the heteroscedasticity and determine the outliers. See Sheather (2009) for discussion on the "bow tie" pattern in the residual plot. The data are in the dataset `bonds`.

7.9.8. Hamilton (1992) presents a dataset concerning the number of accidental oil spills (x) at sea and the amount of oil loss (y) in millions of metric tons for the years 1973–1975. The data are:

	1	2	3	4	5	6	7
x	36.00	48.00	45.00	29.00	49.00	35.00	65.00
y	84.50	67.10	188.00	204.20	213.10	260.50	723.50
	8	9	10	11	12	13	
x	32.00	33.00	9.00	17.00	15.00	8.00	
y	135.60	45.30	1.70	387.80	24.20	15.00	

Hamilton suggests a regression through the origin model for this data.

(a) Obtain the scatterplot y versus x and overlay the the rank-based and LS fits of the regression through the origin models.

(b) Obtain the residual plots and $q-q$ plots of the fits in Part (a). Comment on which fit is better.

(c) Obtain a 95% confidence interval for the slope parameter based on the rank-based fit.

7.9.9. In using a regression through the origin model, one assumes that the intercept parameter is 0. This is a strong assumption. If there is little or no data with x relatively near 0, this is a form of extrapolation and there is little evidence for which to verify the assumption. One simple diagnostic check is to fit an intercept model, also, and then to compare the two fits.

(a) For the data in Exercise 7.9.8, besides the rank-based fit of the regression through the origin model, obtain the rank-based fit of the intercept model. Overlay these fits on the scatterplot of the data. Which, if any, fit is better? Why?

(b) One check for the assumption of a 0 intercept is to use a confidence interval for the intercept as a diagnostic confirmation. Obtain the confidence interval for the rank-based fit.

(c) The fallacy in Part (b), of course, is that it may be a form of extrapolation as discussed at the beginning of this exercise. Is this extrapolation a concern for this problem? Why?

7.9.10. One form of heteroscedasticity that occurs in regression models is when the observations are collected over time and the response varies with time. Good statistical practice dictates plotting the responses and other variables versus time, if time may be a factor. Consider the following data:

	1	2	3	4	5	6	7	8	9	10
x	46	37	34	30	33	24	30	49	54	47
y	132	105	94	71	84	135	10	132	148	132
x	24	33	42	50	47	55	25	44	38	53
y	67	105	104	132	142	130	60	133	204	163

(a) Scatterplot the data. Obtain the the Wilcoxon fit and the residual plot. Do the data appear to be heteroscedastic?

(b) These data were collected over time. For such cases, it is best to plot the residuals versus time. Obtain this plot for our dataset (Wilcoxon residuals versus time, t<-1:20). Comment on the heteroscedasticity.

(c) Assume the response varies directly with time; i.e., $\text{Var}(e_i) = i\sigma^2$, where e_i denotes the random error for the ith case. Appropriate weights in this case are: diag(1/i), where i<-1:20. Use the function wtedrb to fit the data with these weights. Remember to use the design matrix xmat<-cbind(rep(1,20),x), where x is the vector containing the x's.

(d) Compare the precision of fits in parts (a) and (c).

7.9.11. Consider a model where the responses vary directly with time order of the observations:

$$Y_i = 5 - 3x_i + e_i \text{ for } i = 1, \ldots 20$$

where the explanatory variables $(x_1, \ldots, x_{20})$ are generated from a standard normal distribution and the errors $(e_1, \ldots, e_{20})$ are generated from a contaminated normal distribution with $\epsilon = 0.2$ and $\sigma = 9$. Run a simulation comparing the usual Wilcoxon fit with the weighted Wilcoxon fit using the weights `diag(1/t)`. Remember to use the design matrix `xmat<-cbind(rep(1,20),x)` for the weighted Wilcoxon fit. Comment on the simulation, in particular address the following.

  (a) Empirical mean square errors for the estimates of the slope parameter $\beta_1 = -3$.

  (b) Validity of 95% confidence intervals for the slope parameter $\beta_1 = -3$.

7.9.12. Assuming $X \sim SN(\alpha)$ for some $-\infty < \alpha < \infty$, obtain the derivation of expression (7.22).

7.9.13. As in Figure 7.11, obtain the plots of the pdfs and their associated scores for $\alpha = 1, 2, \ldots, 10$. Comment on the trends in the plots as α increases.

7.9.14. Assuming $X \sim SN(\alpha)$ for some $-\infty < \alpha < \infty$, show that the derivative of the optimal scores satisfies expression (7.24). of expression (7.22).

7.9.15. Simulate 50 observations from the model

$$Y_i = 0.01 * x_{i1} + 0.15 * x_{i2} + 0 * x_{i3} + e_i,$$

where x_{i1}, x_{i2} are deviates from a standard normal distribution and $e_i \sim SN(-8)$.

  (a) Obtain the rank-based fit of these data using skew normal scores with $\alpha = -8$. Obtain confidence intervals for the 3 (nonintercept) regression parameters. Did the confidence intervals trap the true values?

  (b) Obtain the Studentized residuals for the fit obtained in Part (a). Using these residuals obtain the residual plot and the normal $q-q$ plot. Comment on the fit.

  (c) Obtain the Wilcoxon fit of these data. What is the estimated precision of fit in Part (a) over the Wilcoxon fit?

7.9.16. On Page 204, Bowerman et al. (2005) present a dataset concerning sales prices of houses in a city in Ohio. The variables are: y is the sale price in \$10,000; x_1 is the total square footage; x_2 is the number of rooms; x_3 is the number of bedrooms; and x_4 is the age of the house at the time data were collected. The sample size is $n = 63$. For the reader's convenience the data are in the dataset `homesales`. Consider the linear model $y = \alpha + \sum_{i=1}^{4} x_i \beta_i + e$.

(a) Use the Hogg-type adaptive scheme discussed in Section 7.6 on these data; i.e., use the function `adaptor`. Which score function did it select?

(b) Comment on the estimated regression coefficients as to significance and what they mean in terms of the problem.

(c) Fit the model using the selected score function in Part (a).

(d) Using the Studentized residuals from Part (b), perform a residual analysis which includes at least a residual plot and a $q-q$ plot. Identify all outliers.

(e) Based on your analysis in (d), what, if any, other models would you fit to these data?

7.9.17. Apply the Hogg-type adaptive scheme of Section 7.6 to data of Exercise 6.5.6. Compare the selected fit with that of the rank-based fits obtained in Exercise 6.5.6. Which of the three is best in terms of precision?

7.9.18. For the data of Exercise 6.5.7, run the adaptive analysis of Section 7.6. Which score function did it select?

7.9.19. Bowerman et al. (2005) discuss the daily viscosity measurements of a manufactured chemical product XB-77-5 for a series of 95 days. Using traditional methods, they determined that the best autoregressive model for this data had order 2. For the reader's convenience, we have placed this data in the dataset `viscosity`.

(a) Plot this time series data.

(b) As in Example 7.8.1, using the Wilcoxon scores, determine the order of the autoregressive series for this data. Use as the maximum order, $p = 4$. Do the results agree with the order determined by Bowerman et al. (2005)?

(c) Obtain the Wilcoxon fit for the autoregressive model using the order determined in the last part. Obtain confidence intervals for the autoregressive parameters.

(d) Write the model expression for the unknown observation at time $t = 96$. Using the fitted model in Part (c), predict the expected viscosity for Day 96.

(e) Determine a confidence interval for the $E(y_{96})$; see Section 4.4.4.

(f) Continue parts (d) and (e) for the observations at times 97 and 98.

7.9.20. For the time series in Exercise 7.9.19, assume that the order of the autoregressive model is 2. Compute the diagnostic TDBETAS between the Wilcoxon and HBR fits. Obtain the plot of the corresponding diagnostic CFITS versus time also. Comment on the diagnostics.

7.9.21. Seber and Wild (1989) present a dataset for the nonlinear model of the form $y = f(x; \boldsymbol{\theta}) + e$, where

$$f(x; \boldsymbol{\theta}) = \frac{\theta_1 x}{\theta_2 + x}.$$

The response y is the enzyme velocity in an enzyme-catalyzed chemical reaction and x denotes the concentration of the substrate. The 12 data points are:

x	2.00	2.00	0.67	0.67	0.40	0.40
y	0.0615	0.0527	0.0334	0.0258	0.0138	0.0258
x	0.29	0.29	0.22	0.22	0.20	0.20
y	0.0129	0.0183	0.0083	0.0169	0.0129	0.0087

(a) Obtain a scatterplot of the data.

(b) Write the R functions for the model and the Jacobian.

(c) Using $\boldsymbol{\theta}_0 = (.1, 1.8)$ as the initial estimate, obtain the Wilcoxon fit of the model. Obtain 95% confidence intervals for θ_1 and θ_2.

(d) Overlay the scatterplot with the fitted model.

(e) Obtain a residual plot and a $q-q$ plot of the residuals. Comment on the quality of the fit.

7.9.22. Obtain the HBR fit of the nonlinear model discussed in Exercise 7.9.21. Then change let $y_1 = y_2 = 0.04$. Obtain both the Wilcoxon and HBR fits for this changed data. Which fit changed less?

7.9.23. The dataset gamnl contains simulated data from the nonlinear model

$$f(x; \boldsymbol{\theta}) = \theta_2^{\theta_1} x^{\theta_1 - 1} e^{-\theta_2 x}.$$

The first and second columns of the data contain the respective x's and y's.

(a) Obtain a scatterplot of the data.

(b) Write the R functions for the model and the Jacobian.

(c) For starting values, the model suggests taking the log of both sides of the model expression and then fit a linear model; but since some of the y values are negative use $log(y + 2)$.

(d) Obtain the Wilcoxon fit of the model and 95% confidence intervals for θ_1 and θ_2.

(e) Overlay the scatterplot with the fitted model.

(f) Obtain a residual plot and a $q-q$ plot of the residuals. Comment on the quality of the fit.

7.9.24. Devore (2012) presents a dataset on the wear life for solid film lubricant. We consider the model

$$Y_i = \frac{\theta_1}{x_{1i}^{\theta_2} x_{2i}^{\theta_3}} + e_i,$$

where Y_i the wear life in hours of a Mil-L-8937-type film, x_{i1} is load in psi, and x_{i2} is the speed of the film in rpm. The data are the result of a 3×3 crossed design with three replicates and are in dataset `wearlife`.

(a) Write the R functions for the model and the Jacobian.

(b) For starting values, as the model suggests, take logs of both sides. Show that $\theta_1 = 36,000$, $\theta_2 = 1.15$, and $\theta_3 = 1.24$ are reasonable starting values.

(c) Obtain the Wilcoxon fit and find 95% confidence intervals for each of the parameters.

(d) Obtain the residual versus fitted values plot and the normal $q-q$ plot of the residuals. Comment on the quality of the fit.

(e) Obtain a scatterplot of x_{i1} versus x_{i2}. There are, of course, 9 treatment combinations. At each combination, plot the predicted wear life based on the fitted model. Obtain 95% confidence intervals for these predictions. Discuss the plot in terms of the model.

8

Cluster Correlated Data

8.1 Introduction

Often in practice, data are collected in clusters. Examples include block designs, repeated measure designs, and designs with random effects. Generally, the observations within a cluster are dependent. Thus the independence assumption of fixed effects linear models breaks down. These models generally include fixed effects, also. Inference (estimation, confidence intervals, and tests of linear hypotheses) for the fixed effects is often of primary importance.

Several rank-based approaches have been considered for analyzing cluster-correlated data. Kloke, McKean, and Rashid (2009) extended the rank-based analysis for linear models discussed in Chapters 4–5 to many cluster models which occur in practice. In their work, the authors, in addition to developing general theory for cluster-correlated data, develop the application of a simple mixed model with one random effect and an arbitrary number of fixed effects and covariates. Kloke and McKean (2011) discuss a rank-based based analysis when the blocks have a compound symmetric variance covariance structure.

In this chapter we illustrate extensions of the rank-based methods discussed in earlier chapters to data which have cluster-correlated responses. For our purpose we consider an experiment done over a number of blocks (clusters) where the observations within a block are correlated. We begin (Section 8.2) by discussing Friedman's nonparametric test for a randomized block design. In Section 8.3, we present the rank-based analysis of Kloke, McKean, and Rashid (2009). Besides tests of general linear hypotheses, this analysis includes estimation with standard errors of fixed effects as well as diagnostic procedures to check the quality of fit. Section 8.4 offers a discussion of robust estimation of variance components. These estimates are also used in the estimation of standard errors and in the Studentized residuals. We end the chapter with a discussion of rank-based procedures for general estimation equation (GEE) models which in terms of assumptions are the most general. Computation by R and R packages of these analyses is highlighted throughout the chapter.

For this chapter we use a common notation which we provide now. Suppose we have m blocks or clusters. Within the kth cluster there are n_k measurements. We may model the ith measurement within the kth cluster as

$$Y_{ki} = \alpha + \boldsymbol{x}_{ki}^T \boldsymbol{\beta} + e_{ki} \text{ for } k = 1, \ldots m, i = 1, \ldots, n_k, \tag{8.1}$$

where x_{ki} is a vector of covariates. The errors between clusters are assumed to be independent and the errors within a block are assumed to be correlated.

At times, dependent data fit into a multivariate frame-work; i.e., a multivariate multiple regression model. In this edition, we have not covered rank-based procedures for multivariate analysis. We refer the reader to Chapter 6 of Hettmansperger and McKean (2011) and Oja (2010) for discussions of these procedures.

8.2 Friedman's Test

The first nonparametric test for cluster-correlated data was developed by Friedman (1937). The goal is to compare the effect of n treatments. Each treatment is applied to each of m experimental units or clusters. In this test a separate ranking is calculated for each of the clusters. The rankings are then averaged for each of the treatments and then compared. If there is a large difference between the average rankings the null hypothesis of no treatment effect is rejected.

Suppose we have n treatments and m clusters each of size n. Suppose all the treatments are randomly assigned once within a cluster. Let Y_{kj} denote the measurement (response) for the jth treatment within cluster (experimental unit) k. Assume the model is

$$Y_{kj} = \alpha + \beta_j + b_k + \epsilon_{kj}, \quad k = 1, \ldots, m, j = 1, \ldots, n, \tag{8.2}$$

where α is an intercept parameter, β_j is the jth treatment effect, b_k is the random effect due to cluster k, and ϵ_{kj} is the $jkth$ random error. Assume that the random errors are iid and are independent of the random effects.

Let R_{kj} denote the rank of Y_{kj} among $Y_{k1}, \ldots, Y_{kn}$. Let

$$\bar{R}_{\cdot j} = \frac{\sum_{k=1}^{m} R_{kj}}{m}.$$

The test statistic is given by

$$T = \frac{12m}{n(n+1)} \sum_{j=1}^{n} \left(\bar{R}_{\cdot j} - \frac{n+1}{2} \right)^2.$$

Under H_0, the test statistic T has an asymptotic χ^2_{n-1} distribution. We illustrate the R computation of Friedman's test with the following example.

Example 8.2.1 (Rounding First Base). This example is discussed in Hollander and Wolfe (1999). In the game of baseball, three methods were evaluated for rounding first base (for an illustration see Figure 7.1 of Hollander and

Wolfe 1999). Label these methods as round out, narrow angle, and wide angle. Each method was evaluated twice for each of $m = 22$ baseball players. The average time of the two runs are in the dataset `firstbase`. Hence, there are 22 blocks (clusters) and one fixed effect (method of base rounding) at three levels. The R function `friedman.test` can take either a numeric data matrix, separate arguments for the response vector, the group vector, and the block vector or a formula.

```
> friedman.test(as.matrix(firstbase))

        Friedman rank sum test

data:  as.matrix(firstbase)
Friedman chi-squared = 11.1429, df = 2, p-value = 0.003805
```

Hence, the difference between methods of rounding first base is significant. Friedman's test is for an overall difference in the methods. Note that it offers no estimate of the effect size between the different methods. Using the rank-based analysis discussed in the next section, we can both test the overall hypothesis and estimate the effect sizes, with standard errors. ∎

8.3   Joint Rankings Estimator

Kloke et al. (2009) showed that rank-based analysis can be extended to cluster-correlated data. In this section we summarize these methods and present examples which illustrate the computation; as we demonstrate the function to determine the fit is `jrfit` and there are, in addition, several of the standard linear model helper functions.

Assume an experiment is done over m blocks or clusters. Note that we use the terms block and cluster interchangeably. Let n_k denote the number of measurements taken within the kth block. Let Y_{ki} denote the response variable for the ith experimental unit within the kth block; let x_{ki} denote the corresponding vector of covariates. Note that the design is general in that x_{ki} may contain, for example, covariates, baseline values, or treatment indicators. The response variable is then modeled as

$$Y_{ki} = \alpha + x_{ki}^T\beta + e_{ki} \text{ for } k = 1,\ldots,m, i = 1,\ldots,n_k, \tag{8.3}$$

where α is the intercept parameter, β is a $p \times 1$ vector of unknown parameters, and e_{ki} is an error term. We assume that the errors within a block are correlated (i.e. e_{ki} & $e_{ki'}$) but the errors between blocks are independent (i.e. e_{ki} & $e_{k'j}$). Further, we assume that e_{ki} has pdf and cdf $f(x)$ and $F(x)$, respectively. Now write model (8.3) in block vector notation as

$$Y_k = \alpha 1_{n_k} + X_k\beta + e_k. \tag{8.4}$$

where $\mathbf{1}_{n_k}$ is an $n_k \times 1$ vector of ones and $\boldsymbol{X}_k = [\boldsymbol{x}_{k1} \ldots \boldsymbol{x}_{kn_k}]^T$ is a $n_k \times p$ design matrix and $\boldsymbol{e}_k = [e_{k1}, \ldots e_{kn_k}]^T$ is a $n_k \times 1$ vector of error terms. Let $N = \sum_{k=1}^m n_k$ denote the total sample size. Let $\boldsymbol{Y} = (\boldsymbol{Y}_1^T, \ldots, \boldsymbol{Y}_m^T)^T$ be the $N \times 1$ vector of all measurements (responses) and consider the matrix formulation of the model as

$$\boldsymbol{Y} = \alpha \mathbf{1}_N + \boldsymbol{X}\boldsymbol{\beta} + \boldsymbol{e} \tag{8.5}$$

where $\mathbf{1}_N$ is an $N \times 1$ vector of ones and $\boldsymbol{X} = [\boldsymbol{X}_1^T \ldots \boldsymbol{X}_m^T]^T$ is a $N \times p$ design matrix and $\boldsymbol{e} = [\boldsymbol{e}_1^T, \ldots \boldsymbol{e}_m^T]^T$ is a $N \times 1$ vector of error terms. Since there is an intercept in the model, we may assume (WLOG) that $\boldsymbol{X}$ is centered.

Select a set of rank scores $a(i) = \varphi[i/(N+1)]$ for a nondecreasing score function φ which is standardized as usual, ($\int \varphi(u)\, du = 0$ and $\int \varphi^2(u)\, du = 1$). As with Rfit, the default score function for jrfit is the Wilcoxon, i.e., $\varphi(u) = \sqrt{12}[u - (1/2)]$. Then the rank-based estimator of $\boldsymbol{\beta}$ is given by

$$\hat{\boldsymbol{\beta}}_\varphi = \text{Argmin}\|\boldsymbol{y} - \boldsymbol{X}\boldsymbol{\beta}\|_\varphi \text{ where } \|\boldsymbol{v}\|_\varphi = \sum_{t=1}^N a(R(v_t))v_t, \quad \boldsymbol{v} \in R^N, \tag{8.6}$$

is Jaeckel's dispersion function.

For formal inference, Kloke et al. (2009) develop the asymptotic distribution of the $\hat{\boldsymbol{\beta}}_\varphi$ under the assumption that the marginal distribution functions of the random vector $\boldsymbol{e}_k$ are the same. This includes two commonly assumed error structures: exchangeable within-block errors as well as the components of $\boldsymbol{e}_k$ following a stationary time series, such as autoregressive of general order. This asymptotic distribution of $\hat{\boldsymbol{\beta}}$ is given by

$$\hat{\boldsymbol{\beta}}_\varphi \dot{\sim} N_p \left(\boldsymbol{\beta}, \tau_\varphi^2 (\boldsymbol{X}^T\boldsymbol{X})^{-1} \left(\sum_{k=1}^m \boldsymbol{X}_k^T \boldsymbol{\Sigma}_{\varphi_k} \boldsymbol{X}_k \right) (\boldsymbol{X}^T\boldsymbol{X})^{-1} \right)$$

where $\boldsymbol{\Sigma}_k = \text{var}(\varphi(F(\boldsymbol{e}_k)))$ and $F(\boldsymbol{e}_k) = [F(e_{k1}), \ldots, F(e_{kn_k})]^T$. To estimate τ_φ, jrfit uses the estimator purposed by Koul et al. (1987).

8.3.1   Estimates of Standard Error

In this section we discuss several approaches to estimating the standard error of the R estimator defined in (8.6). Kloke et al. (2009) develop the inference under the assumption of exchangeable within-block errors; Kloke and McKean (2013) considered two additional estimates and examined the small sample properties of each.

Let $\boldsymbol{V} = \left(\sum_{k=1}^m \boldsymbol{X}_k^T \boldsymbol{\Sigma}_{\varphi_k} \boldsymbol{X}_k \right)$. Let σ_{ij} be the (i,j)th element of $\boldsymbol{\Sigma}_{\varphi k}$. That is $\sigma_{ij} = \text{cov}(\varphi(F(e_{1i})), \varphi(F(e_{1j})))$.

Compound Symmetric

Kloke et al. (2009) discuss estimates of $\boldsymbol{\Sigma}_{\varphi k}$ when the within block errors are exchangeable. Under the assumption of exchangeable errors $\boldsymbol{\Sigma}_{\varphi k}$ reduces to compound symmetric; i.e., $\boldsymbol{\Sigma}_{\varphi k} = [\sigma_{ij}]$ where

$$\sigma_{ij} = \begin{cases} 1 \text{ if } i = j \\ \rho_\varphi \text{ if } i \neq j \end{cases}$$

and $\rho_\varphi = \text{cov}(\varphi(F(e_{11})), \varphi(F(e_{12})))$. An estimate of ρ_φ is

$$\hat{\rho}_\varphi = \frac{1}{M - p} \sum_{k=1}^{m} \sum_{i>j} a(R(\hat{e}_{ki}))a(R(\hat{e}_{kj}))$$

where $M = \sum_{k=1}^{m} \binom{n_k}{2}$.

One advantage of this estimate is that it requires estimation of only one additional parameter. A main disadvantage is that it requires the somewhat strong assumption of exchangeability.

Empirical

A natural estimate of $\boldsymbol{\Sigma}_\varphi$ is the unstructured variance-covariance matrix using the sample correlations. To simplify notation, let $a_{ki} = a(R(\hat{e}_{ki}))$. Estimate σ_{ij} with

$$\hat{\sigma}_{ij} = \sum_{k=1}^{m} (a_{ki} - \bar{a}_{\cdot i})(a_{kj} - \bar{a}_{\cdot j})$$

where $\bar{a}_{\cdot i} = \sum_{k=1}^{m} a_{ki}$.

The advantage of this estimator is that it is general and makes no additional simplifying assumptions. In simulation studies, Kloke and McKean (2013) demonstrate that the sandwich estimator discussed next works at least as well for large samples as this empirical estimate.

Sandwich Estimator

Another natural estimator of $\boldsymbol{V}$ is the sandwich estimator, which for the problem at hand is defined as

$$\frac{m}{m - p} \sum_{k=1}^{m} \boldsymbol{X}_k^T a(R(\hat{e}_k)) a(R(\hat{e}_k))^T \boldsymbol{X}_k.$$

Kloke and McKean (2013) demonstrate that this estimate works well for large samples and should be used when possible. The advantage of this estimator is that it does not require additional assumptions. For very small sample sizes, though, it may lead to biased, often conservative, inference. Simulation studies, however, suggest that when $m \geq 50$ the level is close to α. See Kloke and McKean (2013) for more details. The sandwich estimator is the default in jrfit.

8.3.2   Inference

Simulation studies suggest using t a distribution for tests of hypothesis of the form

$$H_0 : \beta_j = 0 \text{ versus } H_A : \beta_j \neq 0.$$

Specifically, when the standard error (SE) is based on the estimate of the compound symmetric structure, we may reject the null hypothesis at level α provided

$$\left| \frac{\hat{\beta}_j}{\text{SE}(\hat{\beta}_j)} \right| > t_{\alpha, N-p-1-1}.$$

On the other hand, if the sandwich estimator is used, (Section 8.3.1), we test the hypothesis using $df = m$. That is, we reject the null hypothesis at level α if

$$\left| \frac{\hat{\beta}_j}{\text{SE}(\hat{\beta}_j)} \right| > t_{\alpha, m}.$$

These inferences are the default when utilizing the `summary` functions of `jrfit`. We illustrate this discussion with the following examples.

8.3.3   Examples

In this section we present several examples. The first is a simulated example for which we illustrate the package `jrfit`. Following that, we present several real examples.

Simulated Dataset

To fix ideas, we present an analysis of a simulated dataset utilizing both the compound symmetry and sandwich estimators discussed in the previous section.

The setup is as follows:

```
> m<-160  # blocks
> n<-4    # observations per block
> p<-1    # baseline covariate
> k<-2    # trtmnt groups
```

First, we set up the design and simulate a baseline covariate which is normally distributed.

```
> trt<-as.factor(rep(sample(1:k,m,replace=TRUE),each=n))
> block<-rep(1:m,each=n)
> x<-rep(rnorm(m),each=n)
```

Next, we set the overall treatment effect to be $\Delta = 0.5$, so that we can form the response as follows. We simulate the block effects from a t-distribution with 3 degrees of freedom and the random errors from a t-distribution with 5 degrees of freedom. Note that the assumption for exchangeable errors is met.

```
> delta<-0.5
> w<-trt==2
> Z<-model.matrix(~as.factor(block))
> e<-rt(m*n,df=5)
> b<-rt(m,df=3)
> y<-delta*w+Z%*%b+e
```

Note the regression coefficient for the covariate was set to 0.

First we analyze the data with the compound symmetry assumption. The three required arguments to jrfit are the design matrix, the response vector, and the vector denoting block membership. In future releases we plan to incorporate a model statement as we have done in Rfit similar to the one in friedman.test.

```
> library(jrfit)
> X<-cbind(w,x)
> fit<-jrfit(X,y,block,var.type='cs')
> summary(fit)
```

```
Coefficients:
  Estimate Std. Error t-value   p.value
  1.395707   0.165898  8.4130 2.636e-16 ***
w 0.256514   0.252201  1.0171    0.3095
x 0.083595   0.130023  0.6429    0.5205
---
Signif. codes:  0 '***' 0.001 '**' 0.01 '*' 0.05 '.' 0.1 ' ' 1
```

Notice, by default the intercept is displayed in the output. If the inference on the intercept is of interest then set the option int to TRUE in the jrfit summary function. The cell medians model can also be fit as follows.

```
> library(jrfit)
> W<-model.matrix(~trt-1)
> X<-cbind(W,x)
> fit<-jrfit(X,y,block,var.type='cs')
> summary(fit)
```

```
Coefficients:
       Estimate Std. Error t-value   p.value
trt1 1.395707   0.165898  8.4130 2.636e-16 ***
trt2 1.652221   0.168449  9.8085 < 2.2e-16 ***
x    0.083595   0.130023  0.6429    0.5205
---
Signif. codes:  0 '***' 0.001 '**' 0.01 '*' 0.05 '.' 0.1 ' ' 1
```

Next we present the same analysis utilizing the sandwich estimator.

```
> X<-cbind(w,x)
> fit<-jrfit(X,y,block,var.type='sandwich')
> summary(fit)
```

Coefficients:

	Estimate	Std. Error	t-value	p.value	
	1.395707	0.164288	8.4955	1.294e-14	***
w	0.256514	0.247040	1.0383	0.3007	
x	0.083595	0.113422	0.7370	0.4622	

Signif. codes: 0 '***' 0.001 '**' 0.01 '*' 0.05 '.' 0.1 ' ' 1

```
> X<-cbind(W,x)
> fit<-jrfit(X,y,block,var.type='sandwich')
> summary(fit)
```

Coefficients:

	Estimate	Std. Error	t-value	p.value	
trt1	1.395707	0.164288	8.4955	1.294e-14	***
trt2	1.652221	0.166176	9.9426	< 2.2e-16	***
x	0.083595	0.113422	0.7370	0.4622	

Signif. codes: 0 '***' 0.001 '**' 0.01 '*' 0.05 '.' 0.1 ' ' 1

For this example, the results of the analysis based on the compound symmetry method and the analysis based on the sandwich method are quite similar.

Crabgrass Data

Cobb (1998) presented an example of a complete block design concerning the weight of crabgrass. The fixed factors in the experiment were the density of the crabgrass (four levels) and the levels (two) of the three nutrients nitrogen, phosphorus, and potassium. So $p = 6$. Two complete blocks of the experiment were carried out, so altogether there are $N = 64$ observations. In this experiment, block is a random factor. Under each set of experimental conditions, crabgrass was grown in a cup. The response is the dry weight of a unit (cup) of crabgrass, in milligrams. The R analysis of these data were first discussed in Kloke et al. (2009).

The model is a mixed model with one random effect

$$Y_{ki} = \alpha + x_{ki}^T \beta + b_k + \epsilon_{ki} \text{ for } k = 1, 2 \text{ and } j = 1, \ldots, 32.$$

The example below illustrates the rank-based analysis of these data using jrfit.

```
> library(jrfit)
> data(crabgrass)
```

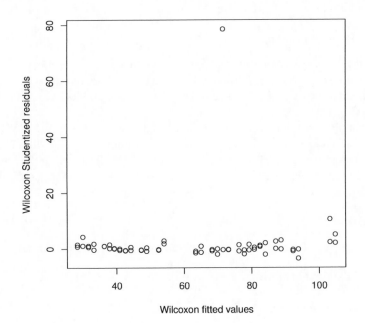

FIGURE 8.1
Plot of fitted values vs. Studentized residuals for crabgrass data.

```
> x<-crabgrass[,1:6]; y<-crabgrass[,7]; block<-crabgrass[,8]
> fit<-jrfit(x,y,block,v1=tcs)
> rm(x,y,block)
> summary(fit)

Coefficients:
   Estimate Std. Error t-value  p.value
   28.31823    2.77923 10.1892 0.009495 **
N  39.87865    3.45475 11.5431 0.007422 **
P  10.96732    4.25586  2.5770 0.123335
K   1.59380    3.91480  0.4071 0.723357
D1 24.08362    1.17606 20.4782 0.002376 **
D2  7.95646    0.50716 15.6882 0.004038 **
D3  3.26657    7.46598  0.4375 0.704443
---
Signif. codes:  0 '***' 0.001 '**' 0.01 '*' 0.05 '.' 0.1 ' ' 1
```

Based on the summary of the fit, the factors nitrogen and density are significant. The Studentized residual plot based on the Wilcoxon fit is given in Figure 8.1. Note the one large outlier in this plot. As discussed in Cobb (1998) and Kloke et al. (2009) this outlier occurred in the data. It impairs the traditional analysis of the data but has little effect on the robust analysis.

Electric Resistance Data

Presented in Stokes et al. (1995), these data are from an experiment to determine if five electrode types performed similarly. Each electrode type (etype) was applied to the arm of 16 subjects. Hence there are 16 blocks and one fixed factor at 5 levels.

The classical nonparametric approach to addressing the question of a difference between the electrode types is to use Friedman's test (Friedman 1937), which is the analysis that Stokes et al. (1995) used. As discussed in Section 8.2, Friedman's test is available in base R via the function `friedman.test`. We illustrate its use with the electrode dataset available in `jrfit`.

```
> library(jrfit)

> friedman.test(resistance~etype|subject,data=eResistance)

        Friedman rank sum test

data:  resistance and etype and subject
Friedman chi-squared = 5.4522, df = 4, p-value = 0.244
```

From the comparison boxplots presented in Figure 8.2 we see there are several outliers in the data.

First we consider a cell medians model where we estimate the median resistance for each type of electrode. There the model is $y_{ki} = \mu_i + b_k + e_{ki}$ where μ_i represents the median resistance for the ith type of electrode, b_k is the kth subject (random) effect, and e_{ki} is the error term encompassing other variability. The variable `etype` is a factor from which we create the design matrix.

```
> x<-model.matrix(~eResistance$etype-1)
> fit<-jrfit(x,eResistance$resistance,eResistance$subject)
> summary(fit)

Coefficients:
                      Estimate Std. Error t-value  p.value
eResistance$etype1     123.998     55.733  2.2248 0.040827 *
eResistance$etype2     211.002     53.894  3.9151 0.001234 **
eResistance$etype3     158.870     56.964  2.7890 0.013137 *
eResistance$etype4     106.526     53.817  1.9794 0.065241 .
eResistance$etype5     109.004     51.219  2.1282 0.049213 *
---
Signif. codes:  0 '***' 0.001 '**' 0.01 '*' 0.05 '.' 0.1 ' ' 1

> muhat<-coef(fit)
> muhat
```

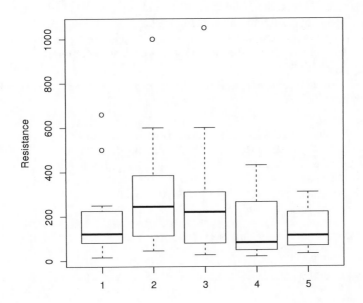

FIGURE 8.2
Comparison boxplots of resistance for five different electrode types.

```
eResistance$etype1 eResistance$etype2 eResistance$etype3
         123.9983           211.0017           158.8703
eResistance$etype4 eResistance$etype5
         106.5256           109.0038
```

As the sample size is small, we will perform our inference assuming compound symmetry. Note, in practice, this can be a strong assumption and may lead to incorrect inference. However, given the nature of this experiment there is likely to be little carry-over effect; hence, we feel comfortable with this assumption.

```
Coefficients:
                         Estimate Std. Error t-value     p.value
eResistance$etype1        123.998     52.674  2.3541  0.0212262 *
eResistance$etype2        211.002     52.674  4.0058  0.0001457 ***
eResistance$etype3        158.870     52.674  3.0161  0.0035080 **
eResistance$etype4        106.526     52.674  2.0223  0.0467555 *
eResistance$etype5        109.004     52.674  2.0694  0.0420014 *
---
Signif. codes:  0 '***' 0.001 '**' 0.01 '*' 0.05 '.' 0.1 ' ' 1
```

Generally, however, we are interested in the effect sizes of the various electrodes. Here the model is $y_{ki} = \alpha + \Delta_i + b_k + e_{ki}$ where Δ_i denotes the effect size. We set $\Delta_1 = 0$ so that the first electrode type is the reference and the others represent median change from the first.

We may estimate the effect sizes directly or calculate them from the estimated cell medians as illustrated in the following code segment.

```
> x<-x[,2:ncol(x)]
> fit<-jrfit(x,eResistance$resistance,eResistance$subject,var.type='cs')
> summary(fit)
```

```
Coefficients:
                    Estimate Std. Error t-value p.value
                    123.994     52.473   2.3630 0.02076 *
eResistance$etype2   87.013     33.567   2.5922 0.01148 *
eResistance$etype3   34.218     33.567   1.0194 0.31134
eResistance$etype4  -17.977     33.567  -0.5356 0.59387
eResistance$etype5  -14.988     33.567  -0.4465 0.65654
---
Signif. codes:  0 '***' 0.001 '**' 0.01 '*' 0.05 '.' 0.1 ' ' 1
```

```
> muhat[2:5]-muhat[1]
```

```
eResistance$etype2 eResistance$etype3 eResistance$etype4 eResistance$etype5
          87.00340           34.87196          -17.47268          -14.99450
```

Next we illustrate a Wald test of the hypothesis that there is no effect due to the different electrodes; i.e.,

$$H_0 : \Delta_i = 0 \text{ for all } k = 2, \ldots, 5 \text{ versus } H_A : \Delta_i \neq 0 \text{ for some } k = 2, \ldots, 5.$$

```
> est<-fit$coef[2:5]
> vest<-fit$varhat[2:5,2:5]
> tstat<-t(est)%*%chol2inv(chol(vest))%*%est/4
> df2<-length(eResistance$resistance)-16-4-1
> pval<-pf(tstat,4,df2,lower.tail=FALSE)
> pval
```

```
            [,1]
[1,] 0.01374127
```

Note that the overall test for effects is highly significant ($p = 0.0137$). This is a much stronger result than that of Friedman's test which was nonsignificant with p-value 0.2440.

8.4   Robust Variance Component Estimators

Consider a cluster-correlated model with a compound symmetry (cs) variance-covariance structure; i.e., a simple mixed model. In many applications, we are interested in estimating the variance components and/or the random effects. For several of the fitting procedures discussed in this chapter, under cs structure, the iterative fitting of the fixed effects depends on estimates of the variance components. Even in the case of the JR fit of Section 8.3, variance components estimates are needed for standard errors and for the Studentization of the residuals. In this section, we discuss a general procedure for the estimation of the variance components and then focus on two procedures, one robust and the other highly efficient.

Consider then the cluster-correlated model (8.3) of the last section. Under cs structure we can write the model as

$$Y_{ki} = \alpha + \boldsymbol{x}_{ki}^T \boldsymbol{\beta} + b_k + \epsilon_{ki}, \quad k = 1, \ldots, m, i = 1, \ldots, n_k, \qquad (8.7)$$

where ϵ_{ki}'s are iid with pdf $f(t)$, b_k's are iid with pdf $g(t)$, and the ϵ_{ki}'s and the b_k's are jointly independent of each other. Hence, the random error for the fixed effects portion of Model (8.7) satisfies $e_{ki} = b_k + \epsilon_{ki}$. Although we could write the following discussion in terms of a general scale parameter (functional) and avoid the assumption of finite variances, for easier interpretation we simply use the variances. The variance components are:

$$
\begin{aligned}
\sigma_b^2 &= \text{Var}(b_k) \\
\sigma_\epsilon^2 &= \text{Var}(\epsilon_{ki}) \\
\sigma_t^2 &= \sigma_b^2 + \sigma_\epsilon^2 \\
\rho &= \frac{\sigma_b^2}{\sigma_b^2 + \sigma_\epsilon^2}
\end{aligned}
\qquad (8.8)
$$

The parameter ρ is often called the **intraclass correlation coefficient**, while the parameter σ_t^2 is often denoted as the **total variance**.

We discuss a general procedure for the estimation of the variance components based on the residuals from a fit of the fixed effects. Let $\hat{\theta}$ and $\hat{\eta}$ be respectively given location and scale estimators. Denote the residuals of the fixed effects fit by $\hat{e}_{ki} = Y_{ki} - \hat{\alpha} - \boldsymbol{x}^T \hat{\boldsymbol{\beta}}$. Then for each cluster $k = 1, \ldots, m$, consider the pseudo model

$$\hat{e}_{ki} = b_k + \epsilon_{ki}, \quad i = 1, \ldots, n_k. \qquad (8.9)$$

Predict b_k by

$$\hat{b}_k = \hat{\theta}(\hat{e}_{k1}, \ldots, \hat{e}_{kn_k}). \qquad (8.10)$$

This estimate is the prediction of the random effect. Then estimate the variance of b_k by the variation of the random effects, i.e.,

$$\hat{\sigma}_b^2 = \hat{\eta}^2(\hat{b}_1, \ldots, \hat{b}_m). \qquad (8.11)$$

For estimation of the variance of ϵ_{ki}, consider Model (8.9), but now move the prediction of the random effect to the left-side; that is, consider the model

$$\hat{e}_{ki} - \hat{b}_k = \hat{\epsilon}_{ki}, \quad i = 1, \ldots, n_k. \qquad (8.12)$$

Then our estimate of the variance of ϵ_{ki} is given by

$$\hat{\sigma}_\epsilon^2 = \hat{\eta}^2(\hat{e}_{11} - \hat{b}_1, \ldots, \hat{e}_{mn_m} - \hat{b}_m).$$ (8.13)

Expressions (8.11) and (8.13) lead to our estimates of total variance given by $\hat{\sigma}_t^2 = \hat{\sigma}_b^2 + \hat{\sigma}_\epsilon^2$ and, hence, our estimate of intraclass correlation coefficient as $\hat{\rho} = \hat{\sigma}_b^2/\hat{\sigma}_t^2$.

Groggel (1983) and Groggel et al. (1988) proposed these estimates of the variance components except the mean is used as the location functional and the sample variance as the scale functional. Assuming that the estimators are consistent estimators for the variance components in the case of iid errors, Groggel showed that they are also consistent for the same parameters when they are based on residuals under certain conditions. Dubnicka (2004) and Kloke et al. (2009) obtained estimates of the variance components using the median and MAD as the respective estimators of location and scale. Since these estimators are consistent for iid errors they are consistent for our model based on the JR fit. The median and MAD comprise our first procedure for variance component estimation and we label it as the MM procedure. As discussed below, we have written R functions which compute these estimates.

Although robust, simulation studies have shown, not surprisingly, that the median and MAD have low efficiency; see Bilgic (2012). A pair of estimators which have shown high efficiency in these studies are the Hodges–Lehmann location estimator and the rank-based dispersion estimator based on Wilcoxon scores. Recall from Chapter 1 that the Hodges–Lehmann location estimator of a sample $X_1, \ldots, X_n$ is the median of the pairwise averages

$$\hat{\theta}_{HL} = \text{med}_{i \leq j}\left\{\frac{X_i + X_j}{2}\right\}.$$ (8.14)

This is the estimator associated with the signed-rank Wilcoxon scores. It is a consistent robust estimator of its functional for asymmetric as well as symmetric error distributions. The associated scale estimator is the dispersion statistic given by

$$\hat{D}(\boldsymbol{X}) = \frac{\sqrt{\pi}}{3n}\sum_{i=1}^{n}\varphi\left[\frac{R(X_i)}{n+1}\right]X_i,$$ (8.15)

where $\varphi(u) = \sqrt{12}[u - (1/2)]$ is the Wilcoxon score function. Note that $D(\boldsymbol{X})$ is just a standardization of the norm, (4.8), of $\boldsymbol{X}$. It is a consistent estimator of its functional for iid random errors as well as residuals; see Chapter 3 of Hettmansperger and McKean (2011). With the multiplicative factor $\sqrt{\pi/3}$ in expression (8.14), $\hat{D}(\boldsymbol{X})$ is a consistent estimator of σ provided that X_i is normally distributed with standard deviation σ. Although more efficient than MAD at the normal model, the statistic $\hat{D}(\boldsymbol{X})$ has an unbounded influence function in the Y-space; see Chapter 3 of Hettmansperger and McKean (2011) for discussion. Hence, it is not robust. We label the procedure based on the Hodges–Lehmann estimator of location and the dispersion function estimator of scale the DHL method.

The R function `vee` computes these variance component estimators for median-MAD (mm) and the HL-disp (dhl) procedures. The input to each consists of the residuals and the vector which identifies the center or cluster. The function returns the vector of variance component estimates and the estimates (predictions) of the random effects. We illustrate their use in the following example.

Example 8.4.1 (Variance Component Estimates). For the example we generated a dataset for a mixed model with a treatment effect (2 levels) and a covariate. The data are over 10 clusters each with a cluster size of 10; i.e., $m = 10$, $n_i \equiv 10$, and $n = 100$. The errors ϵ_{ki} are iid $N(0,1)$ while the random effects are $N(0,3)$. Hence, the intraclass correlation coefficient is $\rho = 3/(1+3) = 0.75$. All fixed effects were set at 0. The following code segment computes the JR fit of the mixed model and the median-MAD and HL-dispersion variance component estimators. For both variance component estimators, we show the estimates $\hat{\sigma}_\epsilon^2$ and $\hat{\sigma}_b^2$.

```
> m<-10   # number of blocks
> n<-10   # number number
> k<-2    # number of treatments
> N<-m*n  # total sample size
> x<-rnorm(N)                           # covariate
> w<-sample(c(0,1),N, replace=TRUE) # treatment indicator
> block<-rep(1:m,n)                     # m blocks of size n
> X<-cbind(x,w)
> Z<-model.matrix(~as.factor(block)-1)
> b<-rnorm(m,sd=3)
> e<-rnorm(N)
> y<-Z%*%b+e
> fit<-jrfit(X,y,block)
> summary(fit)

Coefficients:
    Estimate Std. Error t-value p.value
  -1.959030   2.835237 -0.6910  0.5053
x -0.098897   0.370353 -0.2670  0.7949
w  0.172955   0.766447  0.2257  0.8260

> vee(fit$resid,fit$block,method='mm')

$sigb2
[1] 22.85784

$sige2
[1] 0.6683255

> vee(fit$resid,fit$block)
```

```
$sigb2
         [,1]
[1,] 16.42025
```

```
$sige2
         [,1]
[1,] 0.9971398
```

Exercises 8.7.8–8.7.10 discuss the results of this example and two simulation investigations of the methods median-MAD and HL-dispersion for variance component estimation. ∎

Of the two variance component methods of estimation, due to its robustness, we recommend the median-MAD procedure.

8.5   Multiple Rankings Estimator

A rank-based alternative to using the JR estimator of Section 8.3 is to use the MR estimator developed by Rashid et al. (2012). MR stands for multiple rankings as it utilizes a separate ranking for each cluster; while the JR estimator uses the rankings of the entire dataset or the joint rankings.

The model is the same as 8.3 which we repeat here for reference

$$Y_{ki} = \alpha + \boldsymbol{x}_{ki}^T \boldsymbol{\beta} + e_{ki} \text{ for } k = 1, \ldots, m, i = 1, \ldots, n_k. \tag{8.16}$$

The objective function is the sum of m separate dispersion functions each having a separate ranking given by

$$D(\boldsymbol{\beta}) = \sum_{k=1}^{m} D_k(\boldsymbol{\beta}) \tag{8.17}$$

where $D_k(\boldsymbol{\beta}) = \sum_{i=1}^{n_k} a(R_k(Y_{ki} - \boldsymbol{x}_{ki}^T\boldsymbol{\beta}))(Y_{ki} - \boldsymbol{x}_{ki}^T\boldsymbol{\beta})$ and $R_k(Y_{ki} - \boldsymbol{x}_{ki}^T\boldsymbol{\beta})$ is the ranking of $Y_{ki} - \boldsymbol{x}_{ki}^T\boldsymbol{\beta}$ among $Y_{k1} - \boldsymbol{x}_{k1}^T\boldsymbol{\beta}, \ldots, Y_{kn_k} - \boldsymbol{x}_{kn_k}^T\boldsymbol{\beta}$.

For asymptotic theory of the rank-based fit, we need only assume that the distribution of the random errors have finite Fisher information and that the density is absolutely continuous; see Hettmansperger and McKean (2011: Section 3.4) for discussion. In particular, as with the JR fit, the errors may have an asymmetric distribution or a symmetric distribution. Unlike the JR fit, though, for the MR fit each cluster can have its own score function.

The MR fit obtains the estimates of the fixed effects of the model while it is invariant to the random effects. The invariance of the MR estimate to the random effects is easy to see. Because the rankings are invariant to a constant shift, we have for center k that

$$R_j(Y_{ki} - \alpha - b_k - \boldsymbol{x}_{ki}^T\boldsymbol{\beta}) = R_j(Y_{ki} - \alpha - \boldsymbol{x}_{ki}^T\boldsymbol{\beta}).$$

Because the scores sum to 0, for each center, it follows that the objective function D_{MR}, and thus the MR estimator, are invariant to the random effects.

Rashid et al. (2012) show that the MR estimate is asymptotic normal with mean $\boldsymbol{\beta}$ and variance

$$\tau^2 \boldsymbol{V}_{MR} = \tau^2 \left(\sum_{k=1}^{m} \boldsymbol{X}_k^T \boldsymbol{X}_k \right)^{-1} \tag{8.18}$$

where the scale parameter τ is given by expression (3.19). If Wilcoxon scores are used, this is the usual parameter

$$\tau = \left[\sqrt{12} \int f(x)^2 \, dx \right]^{-1}, \tag{8.19}$$

where $f(x)$ is the pdf of random errors e_{ijl}. In expression (8.19), we are assuming that the same score function is used for each cluster. If this is not the case, letting τ_k denote the scale parameter for cluster k, the asymptotic covariance matrix is $(\sum_{k=1}^{m} \boldsymbol{X}_k^T \boldsymbol{X}_k / \tau_k)^{-1}$.

Model (8.16) assumes that there is no interaction between the center and the fixed effects. Rashid et al. (2012), though, developed a robust test for this interaction based on rank-based estimates which can be used in conjunction with the MR or JR analyses.

Estimation of Scale

A consistent estimator of the scale parameter τ can be obtained as follows. For the kth center, form the vector of residuals $\boldsymbol{r}_{MR}$ with components

$$r_{MR,ki} = y_{ki} - \boldsymbol{x}_{ki}^T \widehat{\boldsymbol{\beta}}_{MR}. \tag{8.20}$$

Denote by $\widehat{\tau}_k$ the estimator of τ proposed by Koul et al. (1987) for each of the m clusters. Note these estimates are invariant to the random effects. Furthermore, it is a consistent estimator of τ. As our estimator of τ, we take the average of these estimators, i.e.,

$$\widehat{\tau}_{MR} = \frac{1}{m} \sum_{k=1}^{m} \widehat{\tau}_j, \tag{8.21}$$

which is consistent for τ. Here we assume that the same score function is used for each cluster. If this is not the case, then, as noted above, each $1/\widehat{\tau}_k$ appears within the sum in expression (8.18).

Inference

Inference based on the MR estimate can be done in the same way as with other linear models discussed in this book. For example, Wald type tests and confidence intervals based on the MR estimates can be formulated in the same way as those based on the JR fit discussed in Section 8.3. Another test statistic, not readily available for the JR procedures, is based on the reduction of dispersion in passing from the reduced to the full model. Denote the reduction in dispersion by

$$RD_{MR} = D_{MR}(\widehat{\boldsymbol{\beta}}_{MR,R}) - D_{MR}(\widehat{\boldsymbol{\beta}}_{MR,F}).\qquad (8.22)$$

Large values of RD_{MR} are indicative of a lack of agreement between the collected data and the null hypothesis. As shown in Rashid et al. (2012), under H_0

$$D^*_{MR} = \frac{RD_{MR}}{\widehat{\tau}_{MR}/2}\ \text{converges in distribution to the}\ \chi^2(q)\ \text{distribution.}\qquad (8.23)$$

A nominal α decision rule is to reject H_0 in favor of H_A, if $D^*_{MR} > \chi^2_\alpha(q)$ where q is the number of constraints.

The drop-in-dispersion test was discussed in Section 4.4.3 and is analogous to the likelihood test statistic $-2\log\Lambda$ in maximum likelihood procedure and has similar interpretation. The use of a measure of dispersion to assess the effectiveness of a model fit to a set of data is common in regression analysis.

The code to compute the MR estimate is in the R package[1] `mrfit`. In the following example, the code segment demonstrates the analysis based on this R function.

Example 8.5.1 (Triglyceride Levels). The dataset `gly4gen` is a simulated dataset similar to an actual trial. Lipid levels for the patients were measured at specified times. The response variable of interest is the change in triglyceride level between the baseline and the week 4 visit. Five treatment groups were considered. The study was conducted at two centers. Centers form the random block effect. Group 1 is referenced.

```
> data(gly4gen)
> X<-with(gly4gen,model.matrix(~as.factor(group)-1))
> X<-X[,2:5]
> y<-gly4gen$diffgly4
> block<-gly4gen$center
> fit<-mrfit(X,y,block,rfit(y~X)$coef[2:5])
> summary(fit)

Coefficients:
                      Estimate  Std. Error  t-ratio
Xas.factor(group)2     0.28523     0.29624   0.96283
Xas.factor(group)3    -2.41176     0.29624  -8.14118
Xas.factor(group)4    33.03831     0.29238 112.99851
Xas.factor(group)5    26.11310     0.29624  88.14797
```

Notice for this simulated data, the triglyceride levels of Groups 3 through 5 differ significantly from Group 1. ∎

---
[1]See https://github.com/kloke/book for more information.

8.6 GEE-Type Estimator

As in the previous sections of this chapter, we consider cluster-correlated data. Using the same notation, let Y_{ki} denote the ith response in the kth cluster, $i = 1, \ldots, n_k$ and $k = 1, \ldots, m$, and let $\boldsymbol{x}_{ki}$ denote the corresponding $p \times 1$ vector of covariates. For the kth cluster, stack the responses and covariates into the respective $n_k \times 1$ vector $\boldsymbol{Y}_k = (Y_{k1}, \ldots, Y_{kn_k})^T$ and $n_k \times p$ matrix $\boldsymbol{X}_k = (\boldsymbol{x}_{k1}^T, \ldots, \boldsymbol{x}_{kn_k}^T)^T$.

In the earlier sections of this chapter, we considered mixed (linear and random) models for $\boldsymbol{Y}_k$. For formal inference, these procedures require that the marginal distributions of the random error vectors for the clusters have the same distribution. In this section, we consider generalized linear models (glm)for cluster-correlated data. For our rank-based procedures, this assumption on the marginal distribution of the random error vector is not required.

Assume that the distribution of Y_{ki} is in the exponential family; that is, the pdf of Y_{ki} is of the form

$$f(y_{ki}) = \exp\{[y_{ki}\theta_{ki} - a(\theta_{ki}) + b(y_{ki})]\phi\}. \tag{8.24}$$

It easily follows that $E[Y_{ki}] = a'(\theta_{ki})$ and $\mathrm{Var}[Y_{ki}] = a''(\theta_{ki})/\phi$. The covariates are included in the model using a specified function h in the following manner

$$\theta_{ki} = h(\boldsymbol{x}_{ki}^T\boldsymbol{\beta}).$$

The function h is called the **link function**. Often the canonical link is used where h is taken to be the identity function; i.e., the covariates are linked to the model via $\theta_{ki} = \boldsymbol{x}_{ki}^T\boldsymbol{\beta}$. The **Hessian** plays an important role in the fitting of the model. For the kth cluster it is the $n_k \times p$ matrix defined by

$$\boldsymbol{D}_k = \frac{\partial a'(\boldsymbol{\theta}_k)}{\partial \boldsymbol{\beta}} = \left[\frac{\partial a'(\theta_{ki})}{\partial \beta_j}\right], \tag{8.25}$$

where $i = 1, \ldots, n_k$, $j = 1, \ldots, p$, and $\boldsymbol{\theta}_k = (\theta_{k1}, \ldots, \theta_{kn_k})^T$.

If the responses within a cluster are independent then the above model is a generalized linear model. We are interested, though, in the cases where there is dependence within a cluster, which often occurs in practice. For the GEE estimates, we do not require the specific covariance of the responses, but, instead, we specify a dependence structure as follows. For cluster k, define the $n_k \times n_k$ matrix $\boldsymbol{V}_k$ by

$$\boldsymbol{V}_k = \boldsymbol{A}_k^{1/2}\boldsymbol{R}_k(\boldsymbol{\alpha})\boldsymbol{A}_k^{1/2}/\phi, \tag{8.26}$$

where $\boldsymbol{A}_k$ is a diagonal matrix with positive elements on the diagonal, $\boldsymbol{R}_k(\boldsymbol{\alpha})$ is a correlation matrix, and $\boldsymbol{\alpha}$ is a vector of parameters. The matrix $\boldsymbol{V}_k$ is called the **working covariance matrix** of $\boldsymbol{Y}_k$, but it need not be the covariance matrix of $\boldsymbol{Y}_k$. For example, in practice, $\boldsymbol{A}_k$ and $\boldsymbol{R}$ are not infrequently taken to be the identity matrices. In this case, we say the covariance structure is **working independence**.

Liang and Zeger (1986) develop an elegant fit of this model based on a set of generalized estimating equations (GEE) which lead to an iterated reweighted least squares (IRLS) solution. As shown by Abebe et al. (2014), each step of their solution minimizes the Euclidean norm for a nonlinear problem. Abebe et al. (2014) developed an analogous rank-based solution that leads to an IRLS robust solution.

Next, we briefly describe Abebe et al.'s solution. Assume that we have selected a score function $\varphi(u)$ which is odd about $1/2$; i.e.,

$$\varphi(1 - u) = -\varphi(u). \tag{8.27}$$

The Wilcoxon score function satisfies this property as do all score functions which are appropriate for symmetric error distributions. As discussed in Remark 8.6.1 this can be easily modified for score functions which do not satisfy (8.27). Suppose further that we have specified the working covariance matrix $\boldsymbol{V}$ and that we also have a consistent estimate $\hat{\boldsymbol{V}}$ of it. Suppose for cluster k that $\hat{\boldsymbol{V}}_k$ is the current estimate of the matrix $\boldsymbol{V}_k$. Let $\boldsymbol{Y}_k^* = \hat{\boldsymbol{V}}_k^{-1/2}\boldsymbol{Y}_k$ and let $g_{ki}(\boldsymbol{\beta}) = \boldsymbol{c}_i^T \mathbf{a}'(\boldsymbol{\theta_k})$, where $\boldsymbol{c}_i^T$ is the ith row of $\hat{\boldsymbol{V}}_k^{-1/2}$. Then the rank-based estimate for the next step minimizes the norm

$$D(\boldsymbol{\beta}) = \sum_{k=1}^{m}\sum_{i=1}^{n_k} \varphi[R(Y_{ki}^* - g_{ki}(\boldsymbol{\beta}))/(n+1)][Y_{ki}^* - g_{ki}(\boldsymbol{\beta})]. \tag{8.28}$$

We next write the rank-based estimator as a weighted LS estimator. Let $e_{ki}(\boldsymbol{\beta}) = Y_{ki}^* - g_{ki}(\boldsymbol{\beta})$ denote the $(k,i)th$ residual and let $m_r(\boldsymbol{\beta}) = \text{med}_{(k,i)}\{e_{ki}(\boldsymbol{\beta})\}$ denote the median of all the residuals. Then, because the scores sum to 0 we have the identity,

$$
\begin{aligned}
D_R(\boldsymbol{\beta}) &= \sum_{k=1}^{m}\sum_{i=1}^{n_k} \varphi[R(e_{ki}(\boldsymbol{\beta}))/(n+1)][e_{ki}(\boldsymbol{\beta}) - m_r(\boldsymbol{\beta})] \\
&= \sum_{rki=1}^{m}\sum_{i=1}^{n_k} \frac{\varphi[R(e_{ki}(\boldsymbol{\beta}))/(n+1)]}{e_{ki}(\boldsymbol{\beta}) - m_r(\boldsymbol{\beta})}[e_{ki}(\boldsymbol{\beta}) - m_r(\boldsymbol{\beta})]^2 \\
&= \sum_{k=1}^{m}\sum_{i=1}^{n_k} w_{ki}(\boldsymbol{\beta})[e_{ki}(\boldsymbol{\beta}) - m_r(\boldsymbol{\beta})]^2 ,
\end{aligned}
\tag{8.29}
$$

where $w_{ki}(\boldsymbol{\beta}) = \varphi[R(e_{ki}(\boldsymbol{\beta}))/(n+1)]/[e_{ki}(\boldsymbol{\beta}) - m_r(\boldsymbol{\beta})]$ is a weight function. We set $w_{ki}(\boldsymbol{\beta})$ to be the maximum of the weights if $e_{ki}(\boldsymbol{\beta}) - m_r(\boldsymbol{\beta}) = 0$. Note that by using the median of the residuals in conjunction with property (8.27) ensures that the weights are positive.

Remark 8.6.1. To accommodate other score functions besides those that satisfy (8.27) quantiles other than the median can easily be used. For example, all rank-based scores are nondecreasing and sum to 0. Hence there are both negative and positive scores. So for a given situation with sample size n, replace the median m_r with the i'th quantile where $a(i') \leq 0$ and $a(j) > 0$ for $j \geq i'$. Then the ensuing weights will be nonnegative. ∎

Expression (8.29) establishes a sequence of IRLS estimates $\left\{\hat{\boldsymbol{\beta}}^{(j)}\right\}$, $j = 1, 2, \ldots$, which satisfy the general estimating equations (GEE) given by

$$\sum_{k=1}^{m} \boldsymbol{D}_k^T \hat{\boldsymbol{V}}_k^{-1/2} \hat{\boldsymbol{W}}_k \hat{\boldsymbol{V}}_k^{-1/2} \left[\boldsymbol{Y}_k - \boldsymbol{a}_k'(\boldsymbol{\theta}) - m_r^* \left(\hat{\boldsymbol{\beta}}^{(j)}\right)\right] = \boldsymbol{0}. \tag{8.30}$$

see Abebe et al. (2014) for details. We refer to these estimates as GEE rank-based estimates, (GEERB).

Also, a Gauss–Newton type algorithm can be developed based on the estimating equations (8.30). Since $\partial \boldsymbol{a}_k'(\boldsymbol{\theta})/\partial \boldsymbol{\beta} = \boldsymbol{D}_k$, a first-order expansion of $\boldsymbol{a}_k'(\boldsymbol{\theta})$ about the jth step estimate $\hat{\boldsymbol{\beta}}^{(j)}$ is

$$\boldsymbol{a}_k'(\boldsymbol{\theta}) = \boldsymbol{a}_k'(\hat{\boldsymbol{\theta}}^{(j)}) + \boldsymbol{D}_K \left(\boldsymbol{\beta} - \hat{\boldsymbol{\beta}}^{(j)}\right).$$

Substituting the right side of this expression for $\boldsymbol{a}_k'(\boldsymbol{\theta})$ in expression (8.30) and solving for $\hat{\boldsymbol{\beta}}^{(j+1)}$ yields

$$\hat{\boldsymbol{\beta}}^{(j+1)} = \hat{\boldsymbol{\beta}}^{(j)} + \left[\sum_{k=1}^{m} \boldsymbol{D}_k^T \hat{\boldsymbol{V}}_k^{-1/2} \hat{\boldsymbol{W}}_k \hat{\boldsymbol{V}}_k^{-1/2} \boldsymbol{D}_k\right]^{-1}$$
$$\times \sum_{k=1}^{m} \boldsymbol{D}_k^T \hat{\boldsymbol{V}}_k^{-1/2} \hat{\boldsymbol{W}}_k \hat{\boldsymbol{V}}_k^{-1/2} \left[\boldsymbol{Y}_k - \boldsymbol{a}_k'(\hat{\boldsymbol{\theta}}^{(j)}) - m_r^* \left(\hat{\boldsymbol{\beta}}^{(j)}\right)\right].$$

Abebe et al. (2014) developed the asymptotic theory for these rank-based GEERB estimates under the assumption of continuous responses. They showed that under regularity conditions, the estimates are asymptotically normal with mean $\boldsymbol{\beta}$ and with the variance-covariance matrix given by

$$\left\{\sum_{k=1}^{m} \boldsymbol{D}_k^T \boldsymbol{V}_k^{-1/2} \boldsymbol{W}_k \boldsymbol{V}_k^{-1/2} \boldsymbol{D}_k\right\}^{-1} \left\{\sum_{k=1}^{m} \boldsymbol{D}_k^T \boldsymbol{V}_k^{-1/2} Var(\boldsymbol{\varphi}_k^{\dagger}) \boldsymbol{V}_k^{-1/2} \boldsymbol{D}_i\right\}$$
$$\times \left\{\sum_{k=1}^{m} \boldsymbol{D}_k^T \boldsymbol{V}_k^{-1/2} \boldsymbol{W}_k \boldsymbol{V}_k^{-1/2} \boldsymbol{D}_k\right\}^{-1}, \tag{8.31}$$

where $\boldsymbol{\varphi}_k^{\dagger}$ denotes the $n_k \times 1$ vector $(\varphi[R(e_{k1}^{\dagger})/(n+1)], \ldots, \varphi[R(e_{kn_k}^{\dagger})/(n+1)])^T$ and $e_{kn_k}^{\dagger}$ is defined by the following expressions:

$$\boldsymbol{Y}_k^{\dagger} = \boldsymbol{V}_k^{-1/2} \boldsymbol{Y}_k = (Y_{k1}^{\dagger}, \ldots, Y_{kn_k}^{\dagger})^T$$
$$\boldsymbol{G}_k^{\dagger}(\boldsymbol{\beta}) = \boldsymbol{V}_k^{-1/2} \boldsymbol{a}_i'(\boldsymbol{\theta}) = [g_{ki}^{\dagger}]$$
$$e_{ki}^{\dagger} = Y_{ki}^{\dagger} - g_{ki}^{\dagger}(\boldsymbol{\beta}). \tag{8.32}$$

A practical implementation of inference the GEERB estimates is discussed in Section 8.6.4. As discussed in Abebe et al. (2014), the GEERB estimates are robust in the Y space (provided the score function is bounded) but not robust in the X space. We are currently developing a HBR GEE estimator which has robustness in both spaces and, also, has high breakdown.

The GEE estimates are quite flexible. The exponential family is a large family of distributions that is often used in practice. The choices of the link functions and working variance-covariance structures allow a large variety of models from which to choose. As shown in expression (8.31), the asymptotic covariance matrix of the estimate takes into account each of these choices; that is, the link function determines the Hessian matrices D_k, (8.25); the working covariance structure determines the matrices V_k, (8.26); and the pdf is reflected in the factor in the middle set of braces. We provided R code to compute GEERB estimates based on the Gauss–Newton step described above. The main driver is the R function `geerfit` which is included in the package[2] `rbgee`. It can be easily modified to compute different options. In the next several subsections, we discuss the weights, link functions, and the working covariance structure, and our R functions associated with these items. We discuss some of the details of R code in the next three subsections which are followed by an illustrative example.

We have only recently developed `geerfit` so we caution the reader that it is an **experimental version** and updates are likely.

8.6.1 Weights

The R function `wtmat(Dmat,eitb,med=TRUE,scores=wscores)` computes the weights, where `eitb` is the vector of current residuals and `Dmat` is the current Hessian matrix. If the option `med` is set to TRUE then it is assumed that the score function is odd about $\frac{1}{2}$ and the median of the current residuals is used in the calculation of the weights as given in expression (8.29). If `med` is FALSE then the percentile discussed in Remark 8.6.1 is used. These are calculated at the current residuals, making use of the Hessian matrix D, (8.25).

8.6.2 Link Function

The link function connects the covariance space to the distribution of the responses. Note that it affects the fitting algorithm in its interaction with the vectors $a_k(\theta_k)$, (8.24), and the Hessian matrices D_k, (8.25). We have set the default link to a linear model. Thus, in the routine `getAp`, for cluster k and with $\beta^{(j)}$ as the current estimate of β, the vector $a_k'(\theta_k)$ is set to $X_k\beta^{(j)}$ and, in the routine `getD`, the matrix D_k is set to X_k. For other link functions these routines have to be changed.

---

[2]See `https://github.com/kloke/book` for more information.

8.6.3 Working Covariance Matrix

The working covariance matrix V, (8.26), is computed in the function `veemat`. Currently there are three options available: working independence, "WI"; compound symmetry (exchangeable), "CS"; and autoregressive order 1, "AR". Default is set at compound symmetry. For the compound symmetry case, this function also sets the method for computing the variance components. Currently, there are two options: the MAD-median option, "MM", and the dispersion and Hodges–Lehmann estimator, "DHL." The default option is the MAD-median option. Recall that the MAD-median option results in robust estimates of the variance components.

8.6.4 Standard Errors

Abebe et al. (2014) performed several Monte Carlo studies of procedures to standardize the GEEBR in terms of validity. One procedure involved estimation of the asymptotic variance-covariance matrix, (8.31), of GEEBR estimators, using the final estimates of the matrices V, W, and D. For cluster k, a simple moment estimator of $Var(\varphi_k^\dagger)$ based on residuals is discussed in Abebe et al. (2014). The resulting estimator of the asymptotic variance-covariance matrix in the Monte Carlo studies, though, appeared to lead to a liberal inference. In the studies, a first-order approximation to the asymptotic variance-covariance matrix appeared to lead to a valid inference. The first-order approximation involves replacing the weight matrix $\hat{W}$ by $\hat{\tau}^{-1}I$, where $\hat{\tau}$ is the estimator of τ, and the matrix $Var(\varphi_k^\dagger)$ by I_k. In our R function the indicator of the variance covariance procedure is the variable `varcovst`. The default setting is `varcovst=="var2"` which results is this approximation while the setting `"var1` results in the estimation of the asymptotic form. The third setting, `"var3"` is a hybrid of the two where just W is approximated. This is similar to a sandwich type estimator.

8.6.5 Examples

The driver of our GEERB fit R function at defaults settings is:

$$geerfit(xmat, y, center, scores = wscores, geemod = "LM",$$
$$structure = "CS", substructure = "MM",$$
$$med = TRUE, varcovst = "var2",$$
$$maxstp = 50, eps = 0.00001, delta = 0.8, hparm = 2).$$

The function `geerfit` assumes that `y` and `xmat` are sorted by center. The HBR version is under development and, hence, currently not available.

The routine returns the estimates of the regression coefficients and their standard errors and t-ratios, along with the variance-covariance estimator and the history (in terms of estimates) of the Newton steps. We illustrate the routine with two examples.

Example 8.6.1. For this example, we simulated data with 5 clusters each of size 10. The covariance structure is compound symmetrical with the variances set at $\sigma_\varepsilon^2 = 1$ and $\sigma_b^2 = 3$, so that the intraclass coefficient is $\rho = 0.75$. The random errors and random effects are normally distributed, $N(0,1)$, with the true β set at $(0.5, 0.35, 0.0)^T$. There are 3 covariates which were generated from a standard normal distribution. The data can be found in the dataset eg1gee. The following R segment loads the data and computes the Wilcoxon GEE fit. We used the default settings for the GEERB fit. In particular, compound symmetry was the assumed covariance structure and the variance component estimates are returned in the code.

```
> xmat<- with(eg1gee,cbind(x1,x2,x3))
> gwfit <- geerfit(xmat,y,block)
> gwfit$tab

          Est         SE      t-ratio
x1   0.54067723 0.1118348   4.83460610
x2  -0.01492029 0.1576557  -0.09463847
x3  -0.20872498 0.1300584  -1.60485548

> vc <- gwfit$vc
> vc

[1] 2.2674708 1.0457868 1.2216840 0.4612129

> rho <- vc[2]/vc[1]
> rho

[1] 0.4612129
```

The GEERB estimates of the three components of β are 0.541, -0.015, and -0.209. Based on the standard error of the estimates, the true value of each component of $\hat{\beta}$ is trapped within the respective 95% confidence interval. The estimates of the variance components are $\hat{\sigma}_t^2 = 2.267$, $\hat{\sigma}_b^2 = 1.046$, $\hat{\sigma}_\varepsilon^2 = 1.222$, and $\hat{\rho} = 0.461$.

We next change the data so that $y_{11} = 53$ instead of 1.53 and rerun the fits:

```
> y[1] <-  53
> gwfit <- geerfit(xmat,y,block)
> gwfit$tab

          Est         SE       t-ratio
x1   0.6069074818 0.1491861   4.068123565
x2  -0.0006604922 0.2119288  -0.003116575
x3  -0.3447768897 0.1454904  -2.369756997

> gwfit$vc
```

```
[1] 2.1359250 0.6595324 1.4763927 0.3087807
```

There is little change in the estimate of β, verifying the robustness of the GEERB estimator. ∎

Example 8.6.2 (Rounding Firtsbase, Continued). Recall that in Example 8.2.1 three methods (round out, narrow angle, and wide angle) for rounding first base were investigated. Twenty-two baseball players served as blocks. The responses are their average times for two replications of each method. Thus the design is a randomized block design. In Example 8.2.1, the data are analyzed using Friedman's test which is significant for treatment effect. Friedman's analysis, though, consists of only a test. In contrast, we next discuss the rank-based analysis based on the GEERB fit. In addition to a test for over all treatment effect, it offers estimates (with standard errors) of size effects, estimates of the variance components, and a residual analysis for checking quality of fit and in determining outliers. We use a design matrix which references the first method. The following code provides the Wald test, based on the fit, which tests for differences among the three methods:

```
> fit <- geerfit(xm,y,center)
> beta <- fit$tab[,1]
> tst <- t(beta)%*%solve(fit$varcov)%*%beta/2
> pv <- 1-pchisq(tst,2)
> c(tst,pv)
```

```
[1] 6.36264299 0.04153074
```

The Wald test is significant at the 5% level. With Friedman's method, this would be the end of the analysis. Let μ_i denote the mean time of method i. The effects of interest are the differences between these means. Because method 1 is referenced, the summary of the fit (fit$tab) provides the inference for $\mu_3 - \mu_1$ and $\mu_2 - \mu_1$. However, the next few lines of code yield the complete inference for comparison of the methods. The summary is displayed in Table 8.1.

```
> h <- matrix(c(-1,1),ncol=2); e32 <- h%*%beta
> se32 <- sqrt(h%*%fit$varcov%*%t(h))
> t32 <- e32/se32; c(e32,se32,t32)
```

```
[1] -0.07826494  0.02271888 -3.44492906
```

Based on the Table 8.1, method 3 significantly differs from the other two methods while methods 1 and 2 do not differ significantly. Hence, overall, method 3 (rounding first base using a wide angle) results in the quickest times. The top panel of Figure 8.3 displays comparison boxplots of the three methods. Outliers are prevalent. The bottom panel shows the $q-q$ plot of the residuals based on the GEERB fit. Notice that three outliers clearly stand

TABLE 8.1

Summary Table of Effects for the Firstbase
Data.

	Effect Est.	SE	t-ratio
mu2 minus mu1	-0.02	0.03	-0.69
mu3 minus mu1	-0.10	0.03	-2.92
mu3 minus mu2	-0.08	0.02	-3.44

out. A simple inspection of the residuals shows that these outliers correspond
to the times of baseball player #22. He appears to be the slowest runner. The
rank-based estimates of the variance components are:

```
> fit$vc
```

```
[1] 0.013402740 0.012364328 0.001038412 0.922522403
```

The estimate of the intraclass correlation coefficient is 0.92, indicating a strong
correlation of running times within players over the three methods. ∎

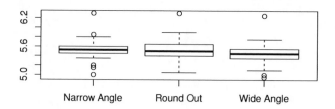

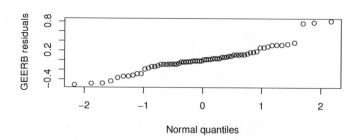

FIGURE 8.3

Comparison boxplots of the three methods and the $q-q$ plot of the residuals
based on the GEERB fit.

8.7 Exercises

8.7.1. Transform the `firstbase` data frame into a vector and create categorical variables for treatment (rounding method) and subject. Obtain the results from `friedman.test` using these data objects.

8.7.2. Referring to Exercise 8.7.1 obtain estimates of the cell medians for each of the rounding methods. Also obtain estimates of the standard errors of these estimates of location using both a compound symmetry estimate as well as the sandwich estimate. Discuss.

8.7.3. It is straightforward to generate data from the simple mixed model. Write a function which generates a sample of size m of blocks each of size n which variance components σ_b and σ_ϵ. Assume normal errors.

8.7.4. Extend 8.7.3 to include errors of a t distribution.

8.7.5. Extend 8.7.3 to allow for different block sizes.

8.7.6. On page 418, Rasmussen (1992) discusses a randomized block design concerned with the readings of four thermometers for melting point of hydroquinone. Three technicians were used as a blocking factor (each technician obtained measurements for all four thermometers). The data are presented next.

MeltPt	Therm.	Tech.	MeltPt	Therm.	Tech.
174.0	1	1	171.5	3	1
173.0	1	2	171.0	3	2
173.5	1	3	173.0	3	3
173.0	2	1	173.5	4	1
172.0	2	2	171.0	4	2
173.0	2	3	172.5	4	3

(a) Let the R vectors `y`, `ind`, `block` contain respectively the melting points, thermometers, and blocks. Argue that the following R code obtains the JR fit of the full model:

```
xmat<-cellx(ind); x2 <-xmat[,2:4]; fit <- jrfit(x2,y,block)
```

(b) Obtain the summary of the above fit. Notice that the first thermometer was referenced. Discuss what the regression coefficients are estimating. Do there seem to be any differences in the thermometers?

(c) Obtain residual and normal q–q plots of the Studentized residuals. Are there any outliers? Discuss the quality of fit.

(d) Argue that the following code obtains the Wald's test of no differences among the parameters. Is this hypothesis rejected at the 5% level?

```
beta<-fit$coef; b<-beta[2:4]; vc <- fit$varhat[2:4,2:4]
tst <- t(b)%*%solve(vc)%*%b/3; pv<-1-pf(tst,3,4)
```

8.7.7. Rasmussen (1992) (page 442) discusses a study on emotions. Each of eight volunteers were requested to express the emotions fear, happiness, depression, and calmness. At the time of expression, their skin potentials in millivolts were measured; hence, there are 32 measurements in all. The data can be found in the dataset `emotion`.

(a) Discuss an appropriate model for the data and obtain a rank-based fit of it.

(b) Using Studentized residuals check quality of fit.

(c) Test to see if there is a significant difference in the skin potential of the four emotions. Conclude at the 5% level.

(d) Obtain a 95% for the shift between the emotions fear and calmness.

(e) Obtain a 95% for the shift between the emotions depression and calmness.

8.7.8. Consider the results of the `jrfit` in Example 8.4.1. Obtain 95% confidence intervals for the fixed effects coefficients. Did they trap the true values?

8.7.9. Run a simulation of size 10,000 on the model simulated in Example 8.4.1. In the simulations, collect the estimates of the variance components σ_b^2 and σ_ϵ^2 for both the median-MAD and the HL-dispersion methods. Obtain the empirical mean square errors. Which method, if any, performed better than the other?

8.7.10. Repeat Exercise 8.7.9, but for this simulation, use the t-distribution with 2 degrees of freedom for both the random errors ϵ_{ki} and the random effects b_k.

Bibliography

Abebe, A., Crimin, K., McKean, J. W., Vidmar, T. J., and Haas, J. V. (2001), "Rank-Based procedures for linear models: Applications to pharmaceutical science data," *Drug Information Journal*, 35, 947–971.

Abebe, A. and McKean, J. W. (2007), "Highly efficient nonlinear regression based on the Wilcoxon norm," in *Festschrift in Honor of Mir Masoom Ali on the Occasion of his Retirement*, ed. Umbach, D., Ball State University, pp. 340–357.

— (2013), "Weighted Wilcoxon estimators in nonlinear regression," *Australian & New Zealand Journal of Statistics*, 55, 401–420.

Abebe, A., McKean, J. W., and Kloke, J. D. (2014), "Iterated reweighted rank-based estimates for GEE models," In preparation.

Adichie, J. N. (1978), "Rank tests of sub-hypotheses in the general linear regression," *The Annals of Statistics*, 6, 1012–1026.

Agresti, A. (1996), *An introduction to categorical analysis*, New York: John Wiley & Sons, Inc.

— (2002), *Categorical data analysis*, New York: John Wiley & Sons, Inc.

Azzalini, A. (1985), "A class of distributions which includes the normal ones," *Scandinavian Journal of Statistics*, 12, 171–178.

— (2014), *The R sn package : The skew-normal and skew-t distributions (version 1.0-0)*, Università di Padova, Italia.

Belsley, D. A., Kuh, K., and Welsch, R. E. (1980), *Regression Diagnostics*, New York: John Wiley & Sons, Inc.

Bilgic, Y. (2012), "Rank-based estimation and prediction for mixed effects models in nested designs," PhD thesis, Western Michigan University, Department of Statistics.

Bowerman, B. L., O'Connell, R. T., and Koehler, A. B. (2005), *Forecasting, time series, and regression: An applied approach*, Australia: Thomson.

Bowman, A. and Azzalini, A. (1997), *Applied smoothing techniques for data analysis: The kernel approach with S-Plus illustrations*, Oxford: Oxford University Press.

Bowman, A. W. and Azzalini, A. (2014), *R package sm: Nonparametric smoothing methods (version 2.2-5.4)*, University of Glasgow, UK and Università di Padova, Italia.

Box, G. E. P., Jenkins, G. M., and Reinsel, G. M. (2008), *Time series analysis: Forecasting and control*, Hoboken, NJ: John Wiley & Sons, Inc., 4th ed.

Breslow, N. E., Day, N. E., et al. (1980), *Statistical methods in cancer research. Vol. 1. The analysis of case-control studies.*, Distributed for IARC by WHO, Geneva, Switzerland.

Canty, A. and Ripley, B. (2013), *boot: Bootstrap R (S-Plus) Functions*, R package version 1.3-9.

Carroll, R. A. and Ruppert, D. (1982), "Robust estimation in heteroscedastic linear models," *The Annals of Statistics*, 10, 424–441.

Chambers, J. M. (2008), *Software for data analysis: Programming with R*, New York: Springer Verlag.

Chang, W. H., McKean, J. W., Naranjo, J. D., and Sheather, S. J. (1999), "High-breakdown rank regression," *Journal of the American Statistical Association*, 205–219.

Chiang, C. Y. and Puri, M. L. (1984), "Rank procedures for testing subhypotheses in linear regression," *Annals of the Institute of Statistical Mathematics*, 36, 35–50.

Cleveland, W., Grosse, E., and Shyu, W. (1992), "Local regression models," in *Statistical Models in S*, eds. Chambers, J. and Hastie, T., Wadsworth & Brooks/Cole, pp. 309–376.

Cobb, G. W. (1998), *Introduction to design and analysis of experiments*, New York: Springer Verlag.

Collett, D. (2003), *Modeling survival data in medical research*, vol. 57, Boca Raton, FL: Chapman & Hall.

Conover, W. J., Johnson, M. E., and Johnson, M. M. (1983), "A comparative study of tests for homogeneity of variances, with applications to the outer continental shelf bidding data," *Technometrics*, 23, 351–361.

Cook, T. D. and DeMets, D. L. (2008), *Introduction to statistical methods for clinical trials*, Boca Raton, FL: Chapman & Hall.

Cox, D. R. (1972), "Regression models and life-tables," *Journal of the Royal Statistical Society. Series B (Methodological)*, 187–220.

Crimin, K., McKean, J. W., and Vidmar, T. J. (2012), "Rank-based estimate of 4-parameter logistic model," *Pharmaceutical Statistics*, 214–221.

Daniel, W. W. (1978), *Applied nonparametric statistics*, Boston: Houghton Mifflin Co.

Davison, A. C. and Hinkley, D. (1997), *Bootstrap methods and their application*, vol. 1, Cambridge: Cambridge University Press.

Devore, J. (2012), *Probability and statistics for engineering and the sciences*, Boston: Brooks/Cole, 8th ed.

Dixon, S. L. and McKean, J. W. (1996), "Rank based analysis of the heteroscedastic linear model," *Journal of the American Statistical Association*, 91, 699–712.

Doksum, K. (2013), "Asymptotic optimality of Hodges–Lehmann inverse rank likelihood estimators," *Statistical Modelling*, 13, 397–407.

Draper, N. L. and Smith, H. (1966), *Applied regression analysis*, New York: John Wiley & Sons, Inc.

Dubnicka, S. R. (2004), "A rank-based estimation procedure for linear models with clustered data," *Journal of Modern Applied Statistics*, 3, 39–48.

Dupont, W. D. (2002), *Statistical modeling for biomedical researchers: A simple introduction to the analysis of complex data*, Cambridge: Cambridge University Press.

Efron, B. and Tibshirani, R. (1993), *An introduction to the bootstrap*, vol. 57, Boca Raton, FL: Chapman & Hall.

Everitt, B. S. and Hothorn, T. (2014), *HSAUR2: A Handbook of Statistical Analyses Using R (2nd Edition)*, R package version 1.1-9.

Faraway, J. (2006), *Extending the linear model with R*, Boca Raton, FL: Chapman & Hall.

Fligner, M. A. and Killeen, T. J. (1976), "Distribution-free two-sample test for scale," *Journal of the American Statistical Association*, 71, 210–213.

Fligner, M. A. and Policello, G. E. (1981), "Robust rank procedures for the Behrens–Fisher problem," *Journal of the American Statistical Association*, 76, 162–168.

Fox, J. and Weisberg, S. (2011), *An R and S-PLUS Companion to Applied Regression, 2nd Edition*, Thousand Oaks, CA: SAGE, chap. Bootstrapping Regression Models.

Friedman, M. (1937), "The use of ranks to avoid the assumption of normality implicit in the analysis of variance," *Journal of the American Statistical Association*, 32, 675–701.

Graybill, F. A. (1976), *Theory and application of the linear model*, North Scituate, Ma: Duxbury Press.

Graybill, F. A. and Iyer, H. K. (1994), *Regression analysis: Concepts and applications*, Belmont, CA: Duxberry Press.

Greenwood, M. (1926), "The errors of sampling of the survivorship tables," *Reports on Public Health and Statistical Subjects*, 33, 26.

Groggel, D. J. (1983), "Asymptotic nonparametric confidence intervals for the ratio of scale parameters in balanced one-way random effects models," PhD thesis, University of Florida, Department of Statistics.

Groggel, D. J., Eackerly, D., and Rao, P. (1988), "Nonparametric estimation in one-way random effects models," *Communications in Statistics: Simulation and Computation*, 17, 887–903.

Hájek, J. and Šidák, Z. (1967), *Theory of rank tests*, New York: Academic Press.

Hamilton, L. C. (1992), *Regression with graphics*, Pacific Grove, CA: Brooks/Cole.

Hawkins, D. M., Bradu, D., and Kass, G. V. (1984), "Location of several outliers in multiple regression data using elemental sets," *Technometrics*, 26, 197–208.

Hettmansperger, T. P. and McKean, J. W. (1973), "On testing for significant change in CxC tables," *Communications in Statistics*, 2, 551–560.

— (2011), *Robust nonparametric statistical methods, 2nd Edition*, Boca Raton, FL: Chapman & Hall.

Higgins, J. J. (2003), *Introduction to modern nonparametric statistics*, Duxbury Press.

Hocking, R. R. (1985), *The analysis of linear models*, Brooks/Cole Publishing Company Monteray, California.

Hodges, J. L. and Lehmann, E. L. (1962), "Rank methods for combination of independent experiments in analysis of variance," *The Annals of Mathematical Statistics*, 33, 482–497.

— (1963), "Estimation of location based on rank tests," *The Annals of Mathematical Statistics*, 34, 598–611.

Hogg, R. V., McKean, J. W., and Craig, A. T. (2013), *Introduction to mathematical statistics, 7th Ed.*, Boston: Pearson.

Hollander, M. and Wolfe, D. A. (1999), *Nonparametric statistical methods, 2nd Edition*, New York: John Wiley & Sons, Inc.

Huitema, B. E. (2011), *The analysis of covariance and alternatives, 2nd ed.*, New York: John Wiley & Sons, Inc.

Jaeckel, L. A. (1972), "Estimating regression coefficients by minimizing the dispersion of residuals," *The Annals of Mathematical Statistics*, 43, 1449–1458.

Jin, Z., Lin, D., Wei, L., and Ying, Z. (2003), "Rank-based inference for the accelerated failure time model," *Biometrika*, 90, 341–353.

Johnson, B. A. and Peng, L. (2008), "Rank-based variable selection," *Journal of Nonparametric Statistics*, 20, 241–252.

Jurečková, J. (1971), "Nonparametric estimate of regression coefficients," *The Annals of Mathematical Statistics*, 42, 1328–1338.

Kalbfleisch, J. D. and Prentice, R. L. (2002), *The statistical analysis of failure time data*, New York: John Wiley & Sons, Inc., 2nd ed.

Kapenga, J. A. and McKean, J. W. (1989), "Spline estimation of the optimal score function for linear models," in *American Statistical Association: Proceedings of the Statistical Computing Section*, pp. 227–232.

Kaplan, E. L. and Meier, P. (1958), "Nonparametric estimation from incomplete observations," *Journal of the American Statistical Association*, 53, 457–481.

Kendall, M. and Stuart, A. (1979), *The advanced theory of statistics*, New York: Macmillan.

Kloke, J. and Cook, T. (2014), "Nonparametric, covariate-adjusted hypothesis tests using R estimation for clinical trials," In preparation.

Kloke, J., McKean, J., Kimes, P., and Parker, H. (2010), "Adaptive nonparametric statistics with applications to gene expression data," http://www.biostat.wisc.edu/~kloke/useR2010.pdf.

Kloke, J. D. and McKean, J. W. (2011), "Rank-based estimation for Arnold transformed data," in *Nonparametric Statistics and Mixture Models: A Festschrift in Honor of Thomas P. Hettmansperger*, eds. Hunter, D. R., Richards, D. P., and Rosenberger, J. L., Word Press, pp. 183–203.

— (2012), "Rfit: Rank-based estimation for linear models," *The R Journal*, 4, 57–64.

— (2013), "Small sample properties of JR estimators," in *JSM Proceedings*, Alexandria, VA: American Statistical Association.

Kloke, J. D., McKean, J. W., and Rashid, M. (2009), "Rank-based estimation and associated inferences for linear models with cluster correlated errors," *Journal of the American Statistical Association*, 104, 384–390.

Koenker, R. (2013), *quantreg: Quantile Regression*, R package version 5.05.

Koul, H. L. and Saleh, A. K. M. E. (1993), "R-estimation of the parameters of autoregressive [AR(p)] models," *The Annals of Statistics*, 21, 534–551.

Koul, H. L., Sievers, G. L., and McKean, J. W. (1987), "An estimator of the scale parameter for the rank analysis of linear models under general score functions," *Scandinavian Journal of Statistics*, 14, 131–141.

Kruskal, W. and Wallis, W. (1952), "Use of ranks in one-criterion variance analysis," *Journal of the American Statistical Association*, 47, 583–621.

Leisch, F. (2002), "Sweave: Dynamic generation of statistical reports using literate data analysis," in *Compstat 2002 — Proceedings in Computational Statistics*, eds. Härdle, W. and Rönz, B., Physica Verlag, Heidelberg, pp. 575–580, iSBN 3-7908-1517-9.

Liang, K. Y. and Zeger, S. L. (1986), "Longitudinal data analysis using generalized linear models," *Biometrika*, 73, 13–22.

McKean, J. and Sheather, S. (1991), "Small sample properties of robust analyses of linear models based on R-estimates: A survey," in *Directions in Robust Statistics and Diagnostics, Part II*, eds. Stahel, W. and Weisberg, S., New York: Springer Verlag, pp. 1–19.

McKean, J. W., Huitema, B. E., and Naranjo, J. D. (2001), "A robust method for the analysis of experiments with ordered treatment levels," *Psychological Reports*, 89, 267–273.

McKean, J. W. and Kloke, J. D. (2014), "Efficent and adaptive rank-based fits for linear models with skew-normal errors," *Journal of Statistical Distributions and Applications*, in press.

McKean, J. W., Naranjo, J. D., and Sheather, S. J. (1996a), "Diagnostics to detect differences in robust fits of linear models," *Computational Statistics*, 11, 223–243.

McKean, J. W. and Schrader, R. (1980), "The geometry of robust procedures in linear models," *Journal of the Royal Statistical Society, Series B, Methodological*, 42, 366–371.

McKean, J. W. and Sheather, S. J. (2009), "Diagnostic procedures," *Wiley Interdisciplinary Reviews: Computational Statistics*, 1(2), 221–233.

McKean, J. W., Sheather, S. J., and Hettmansperger, T. P. (1994), "Robust and high breakdown fits of polynomial models," *Technometrics*, 36, 409–415.

McKean, J. W. and Sievers, G. L. (1989), "Rank scores suitable for analysis of linear models under asymmetric error distributions," *Technometrics*, 31, 207–218.

McKean, J. W., Vidmar, T. J., and Sievers, G. (1989), "A robust two-stage multiple comparison procedure with application to a random drug screen," *Biometrics*, 45, 1281–1297.

Miliken, G. A. and Johnson, D. E. (1984), *Analysis of messy data, Volume 1*, New York: Van Nostrand Reinhold Company.

Morrison, D. F. (1983), *Applied linear statistical models*, Englewood Cliffs, New Jersey: Prentice Hall.

Naranjo, J. D. and McKean, J. W. (1997), "Rank regression with estimated scores," *Statistics and Probability Letters*, 33, 209–216.

Oja, H. (2010), *Multivariate nonparametric methods with R*, New York: Springer.

Okyere, G. (2011), "Robust adaptive schemes for linear mixed models," PhD thesis, Western Michigan University, Department of Statistics.

Puri, M. L. and Sen, P. K. (1985), *Nonparametric methods in general linear models*, New York: John Wiley & Sons, Inc.

R Development Core Team (2010), *R: A language and environment for statistical computing*, R Foundation for Statistical Computing, Vienna, Austria, ISBN 3-900051-07-0.

Rashid, M. M., McKean, J. W., and Kloke, J. D. (2012), "R estimates and associated inferences for mixed models with covariates in a multicenter clincal trial," *Statistics in Biopharmaceutical Research*, 4, 37–49.

Rasmussen, S. (1992), *An introduction to statistics with data analysis*, Pacific Grove, CA: Brooks/Cole Publishing Co.

Rousseeuw, P. and Van Driessen, K. (1999), "A fast algorithm for the minimum covariance determinant estimator," *Technometrics*, 41, 212–223.

Rousseeuw, P. J., Leroy, A. M., and Wiley, J. (1987), *Robust regression and outlier detection*, vol. 3, Wiley Online Library.

Rousseeuw, P. J. and van Zomeren, B. C. (1990), "Unmasking multivariate outliers and leverage points," *Journal of the American Statistical Association*, 85, 633–648.

Seber, G. A. F. and Wild, C. J. (1989), *Nonlinear regression*, New York: John Wiley & Sons, Inc.

Sheather, S. J. (2009), *A modern approach to regression with R*, New York: Springer.

Shomrani, A. (2003), "A Comparison of Different Schemes for Selecting and Estimating Score Functions Based on Residuals," PhD thesis, Western Michigan University, Department of Statistics.

Siegel, S. (1956), *Nonparametric statistics*, New York: McGraw-Hill.

Stokes, M. E., Davis, M. E. S. C. S., Koch, G. G., and Davis, C. S. (1995), *Categorical data analysis using the SAS system*, Cary, NC: SAS institute.

Terpstra, J., McKean, J. W., and Anderson, K. (2003), "Studentized autoregressive time series residuals," *Computational Statistics*, 18, 123–141.

Terpstra, J., McKean, J. W., and Naranjo, J. D. (2000), "Highly efficient weighted Wilcoxon estimates for autoregression," *Statistics*, 35, 45–80.

— (2001), "GR-Estimates for an autoregressive time series," *Statistics and Probability Letters*, 51, 165–172–80.

Terpstra, J. F. and McKean, J. W. (2005), "Rank-based analyses of linear models using R," *Journal of Statistical Software*, 14, 1–26.

Theil, H. (1950), "A rank-invariant method of linear and polynomial regression analysis, III," *Proc. Kon. Ned. Akad. v. Wetensch*, 53, 1397–1412.

Therneau, T. (2013), *A Package for Survival Analysis in S*, R package version 2.37-4.

Therneau, T. M. and Grambsch, P. M. (2000), *Modeling survival data*, New York: Springer Verlag.

Trautmann, H., Steuer, D., Mersmann, O., and Bornkamp, B. (2014), *truncnorm: Truncated normal distribution*, R package version 1.0-7.

Tryon, P. V. and Hettmansperger, T. P. (1973), "A class of non-parametric tests for homogeneity against ordered alternatives," *The Annals of Statistics*, 1, 1061–1070.

Venables, W. N. and Ripley, B. D. (2002), *Modern applied statistics with S*, New York: Springer Verlag.

Wahba, G. (1990), *Spline models for observational data*, vol. 59, Philadelphia: Society for Industrial and Applied Mathematics.

Watcharotone, K. (2010), "On Robustification of Some Procedures Used in Analysis of Covariance," PhD thesis, Western Michigan University, Department of Statistics.

Wells, J. M. and Wells, M. A. (1967), "Note on project SCUD," *Proceedings 5th Berkeley Symposium*, V, 357–369.

Wickham, H. (2009), *ggplot2: elegant graphics for data analysis*, New York: Springer.

— (2011), "The split-apply-combine strategy for data analysis," *Journal of Statistical Software*, 40, 1–29.

Wood, S. (2006), *Generalized additive models: An introduction with R*, Boca Raton, FL: Chapman & Hall.

Index